Benjamin Loewy

A graduate Course of Natural Science - Experimental and Theoretical for Schools and Colleges

Part. II- Second and Third Year's Course

Benjamin Loewy

A graduate Course of Natural Science - Experimental and Theoretical for Schools and Colleges
Part. II- Second and Third Year's Course

ISBN/EAN: 9783337026523

Printed in Europe, USA, Canada, Australia, Japan

Cover: Foto ©berggeist007 / pixelio.de

More available books at **www.hansebooks.com**

A GRADUATED COURSE

OF

NATURAL SCIENCE

EXPERIMENTAL AND THEORETICAL

FOR

SCHOOLS AND COLLEGES

PART II.—SECOND AND THIRD YEAR'S COURSE

For the Intermediate Classes of Colleges
and Technical Schools

BY

BENJAMIN LOEWY, F.R.A.S., ETC.

EXAMINER IN EXPERIMENTAL PHYSICS TO THE COLLEGE OF PRECEPTORS, LONDON

WITH SIXTY DIAGRAMS

London

MACMILLAN AND CO.

AND NEW YORK

1891

PREFACE

THIS second part will, I hope, be found well adapted to the requirements of the growing number of middle class and technical schools in which a general course of experimental and theoretical work is desired, which is chiefly directed to fundamental facts, and those principles of Physics and Chemistry which have found the widest practical applications. It is now generally recognised that such an introductory course should precede more special studies, and should lay a sound foundation for professional and more advanced practical work. I have throughout aimed at rendering the experiments feasible with a very limited apparatus, and inexpensive materials and appliances. The Appendix, giving "Hints" for the practical work, has received special care, and I trust that no beginner will in vain turn to it for advice in any difficulty.

I have again to express my great obligation to Mr. L. R. Wilberforce, Principal Demonstrator at the Cavendish

Laboratory, Cambridge, for his careful revision of the whole, and for enriching both the experimental and the theoretical part by many valuable additions and alterations which his practical experience as a teacher suggested, to the great advantage of the book.

THE WOODLANDS, ISLEWORTH,
September 1891.

TABLE OF CONTENTS

CHAPTER		PAGE
I.	Matter, Mass, Force, Gravity, Elasticity	1
II.	Work done by Forces—Friction	10
III.	Inertia—Work—Energy	18
IV.	Centre of Gravity, or Centre of Mass—Equilibrium	27
V.	The Lever	39
VI.	The Pulley	52
VII.	The Inclined Plane—The Screw—The Wedge	62
VIII.	Momentum	72
IX.	Energy	82
X.	The Pendulum	95
XI.	Sound a form of Energy	107
XII.	Waves of Sound—Pitch—Air Columns as Resonators	115
XIII.	Radiant Energy	124
XIV.	Reflection	135
XV.	Refraction	146
XVI.	Dispersion of Light	156

CONTENTS

CHAPTER		PAGE
XVII.	CHEMICAL ACTION OF RADIANT ENERGY	167
XVIII.	ELECTRICAL ENERGY	174
XIX.	EFFECTS OF CURRENT ELECTRICITY	182
XX.	FURTHER EFFECTS OF CURRENT ELECTRICITY	190
XXI.	CHEMICAL EFFECTS OF ELECTRICAL CURRENTS	199
XXII.	CHEMICAL EFFECTS OF ELECTRICAL CURRENTS (*Continued*)	207
XXIII.	CHEMICAL ACTION BETWEEN METALS AND ACIDS	214
XXIV.	DEGREES OF CHEMICAL ACTION. AFFINITY. CHEMICAL ENERGY	222

APPENDIX

HINTS FOR PERFORMING THE EXPERIMENTS . . 231

CHAPTER I

MATTER, MASS, FORCE, GRAVITY, ELASTICITY

Experiment 1.—Tie the ends of two fine threads, each about a foot long, to weights of 1 lb. and ½ lb., and place the two bodies before you on the table.

You can see and feel them; they also offer resistance when you try to break them or to alter their shape. They are, therefore, both solid matter, and since both are iron, or both brass, as the case may be, they consist of the same kind of matter. If one were iron and the other brass, we should say that they consisted of different kinds of matter. Most probably their shapes, or external outlines, are similar to one another. There is, however, one essential difference between them: their bulk or volume is not the same. One contains more matter than the other; we should have to add matter, either brass or iron, to the one having the smaller bulk, if we wish them both to be alike as regards their volume. Now, as the quantity of matter in a body is, for the sake of shortness, commonly called the *mass* of a body, we may express the difference between the two bodies before us by saying that they differ in this, that one has a greater mass than the other. Seeing that they consist of the same kind of matter, and that one has half the bulk of the other, we may say that the mass of this smaller one is one-half of that of the larger one.

We shall, however, see afterwards that the term *mass of a body* means something more than what is stated above.

Experiment 2.—Take the free end of each thread between the thumb and the forefinger of each hand, and raise both hands until the two bodies hang freely a few inches above the table.

Observe, while thus holding up both masses, that mutual action is taking place between each of your hands and the body which it holds, and that this action is *transmitted* or sent through the strings from both ends at the same time. The two bodies appear to pull so as to stretch the threads, which without this pull would hang loosely; you feel also that an effort is required on your part to counteract this pull. Further, you feel that this effort is not equal for both hands; it is greater in the case of the larger mass than in that of the smaller.

Observe further that both threads form a straight line, and appear to have the same direction, that is, they are parallel to each other. If we consider our earth as a perfect globe, this straight line if continued would pass through its centre. From this it would follow that the directions of both threads would pass through the same point, and that in reality they are not parallel. But consider how far the centre of the earth is from where you stand, nearly 4000 miles, and how short the threads are compared with the distance at which their directions cut one another. We may, therefore, say without sensible error that they are parallel.

The direction of the threads in this experiment is called the *vertical* or *perpendicular* direction; that of a line at right angles to it is called the *horizontal* direction.

Experiment 3.—Put aside the thread with the smaller mass, and, while holding the larger one as before, lift it vertically upwards through about 8 or 10 inches with the free hand and then let it go.

We can tell beforehand, from previous everyday experience, what will happen. The body immediately falls from the point at which the support of your hand was withdrawn as far as the attached thread will allow, and this downward motion takes place in the same direction as that previously shown by the stretched thread. We know also from previous experience what would happen if we did not hold the other end of the thread; the body would go on falling until it reached the table; and if it were not stopped by the resistance of the solid table it would not cease falling until it reached the ground. Thus while we were holding up the body there was, besides the action observed between it and the hand, mutual action between the body and the earth. It is this action which causes the body to move as near to the ground as possible when allowed, and which causes the string to be stretched while the body is suspended by it.

We ought to observe also in this experiment that the body, when reaching the lowest point to which the string will allow it to fall, exerts a stronger action upon the string and the hand than previously while hanging at rest; it causes a sudden jerk at the end of the string, and an additional effort will be required by our hand to prevent its further fall.

Experiment 4.—Throw an ordinary marble gently up into the air so that it may rise 3 or 4 feet above the table.

Observe carefully the motion of the body. It will move upwards, but more and more slowly until it reaches the highest point to which it rises; it will then appear to be at rest for an extremely short interval of time, and then begin to move downwards, first slowly, then more quickly, until it reaches the table. Here it will rebound and rise again to a height of a few inches, fall again, rise again, though much less than before, and finally come to rest on the table.

The preceding experiments have shown us several bodies in various states as regards *rest* and *motion*. All were at rest upon the table before they were objects of our experiments. They were then set in motion by our hands or by means of strings; one of the weights was raised and allowed to fall; the marble was thrown up, and after moving downwards again was seen to rise and fall again before coming to rest once more upon the table. Now when a body at rest begins to move, or a body in motion comes to rest, or changes either the direction or the rate at which its motion proceeds, that is, if while in motion it is observed gradually to move more quickly or more slowly than before, we naturally say that there must be some cause for this change in the behaviour of the body. Experience shows us in many cases the immediate cause of such changes; in other cases the cause is not so obvious, and it requires careful investigation by experiment and observation to discover it. In any case we are led to recognise the fact that the various changes which we see constantly going on around us must be ascribed to *different* causes. Now every cause which changes, or tends to change, the state of rest or motion of a body is called a *force*.

Experiment 5.—**Throw up a ball as before, but instead of allowing it to fall upon the table catch it with your hands.**

Here we caused a body to move which was previously at rest, and we stopped its motion again; in both cases by an effort of our own body. The cause which produced and stopped motion here cannot be the same as that which made the marble rise again after it had reached the table and then fall, in Experiment 4. Hence we give to the force applied to the ball in this experiment a special name, and call it *muscular force*. It is this force which is employed whenever we support or lift a body, pull, or push, or press it in any way.

On the other hand, the cause which made the body fall in Experiment 3 when our hand was withdrawn from it, cannot be muscular force. It is called the *force of gravity* (from the Latin *gravis*, heavy), and is the cause to which the action observed between the earth and the suspended bodies in Experiment 2 is due. Innumerable instances of the action of this force occur in our daily experience. Slates or bricks torn by the wind from roofs or walls, rain, snow, and hail, fall from a higher level to the earth; immense masses of water rush downwards over precipices as waterfalls; the weights attached to the wheels of a clock, as well as the pendulum, are moved by gravity; the heavy lead of the navigator sinks through the water of the ocean down to the bottom and measures the depth of the sea. These observations show that all bodies on the earth are constantly subject to the action of this force, no matter what the distance between the earth and the bodies; and we may thus state it as a *law* of gravity, that all terrestrial bodies move towards the centre of the earth, unless prevented from doing so by some other force.

We have also seen in Experiment 2 that a smaller mass required less muscular force than a larger mass to support it against the action of gravity, that is, to prevent it from falling. It follows from this that the action of this force is less upon a smaller mass than upon a larger. But the action of gravity upon the mass of a material body is for shortness' sake called the *weight* of that body; hence we may say, the larger the mass of a body the greater is its weight, the smaller a mass the less is its weight.

We must, however, be very careful to distinguish between the mass and the weight of a given body. The quantity of matter in it cannot be the same thing as the action of gravity upon it. The mass of a body always remains the same unless some matter is added to it or taken from it. If the force of gravity were always the same at every place upon the earth the weight of a body would also always be the same; but this force is subject to certain variations. Thus our mass which was denoted as representing 1 lb. is a definite quantity of brass or iron in whatever place it may be; but it would require a little more muscular force to hold it up against gravity if you would make the experiment nearer to the surface of the earth than where you are now; for example, if you were at the level of the sea, or if you were to take the body to a country nearer to either pole of the earth than where you are; again, it would press less heavily upon your hand if you could go to the equator, or up to a considerable height above the earth's surface. Some of these experiments have been made, and it has been found that the force of gravity is not the same at every place on the earth's surface.

In Experiment 4 we saw the marble, after it had fallen upon the table, move upwards several times and fall down again, before coming to rest. The cause of the downward motion we now know is called gravity. The upward motion must clearly be due to another cause; this cause is called the force of *elasticity*.

Experiment 6.—Throw up a ball of moist clay, and also one of india-rubber, one after the other, to about the same height as the marble in Experiment 4.

Both will fall down after reaching the highest point just as the marble did, but the ball of clay will not rebound at all, while the india-rubber ball will rise much higher than the marble did in the former experiment. On the other hand, the clay shows us better than the other bodies what happened when the falling body reached the table: the substance of the ball is compressed by the collision with the table, and the round surface of the ball has become flattened where it touched the table. No doubt the flat surface of the table suffers a similar compression where the ball comes into contact with it, though we cannot see it. If we press the india-rubber ball with the finger at some point, we see that it is flattened where our muscular force is applied; on removing the finger the pressed part begins to move at once, and the original shape of the ball is restored. Hence there must be force acting within bodies which tends to restore their form when it is altered by the action of other forces. This force is called *elasticity*, and we see from our experiments that the force acts very little or not perceptibly in a moist clay ball, but more strongly in a marble, or in an india-rubber ball. Bodies in which this force produces more striking effects than in other substances, are called *elastic bodies*; steel, ivory, glass, are examples of such substances.

QUESTIONS ON CHAPTER I

1. What do we commonly understand when we speak of the *mass* of a body?

2. Make a list of bodies which have a very large mass; and one of bodies which have a very small mass.

3. Could you by merely looking at two bodies say which has the greater mass? In what cases could you decide the question in that way?

4. What is meant by the *vertical* or *perpendicular* direction on the earth's surface? What by the *horizontal* direction? What is a plumb-line? What is it used for?

5. A man near the edge of a precipice which is some 5000 feet high lets down two plumb-lines, which reach nearly to the foot of it. At the top they are exactly 50 inches apart; is there any reason to think that they will be the same distance apart near the ground below, or not?

6. Mention bodies which you consider as being usually at rest; and mention others which are usually in motion; lastly, mention some that you have seen at rest or in motion at one time, and in the opposite state at another time, and state what you consider the cause of the change in each case.

7. Mention instances in which you have seen bodies in motion change either the direction, or the rate of motion (the "pace"); and state what you consider the cause of the change in each case.

8. What do we mean when we speak of a *force*? Give examples of forces, and describe experiments which show that those you have mentioned agree with your explanation of the term force.

9. Compare the action of gravity as a force with that of

elasticity, and that of muscular force, stating all points of agreement and difference in their modes of action. Which of the three forces mentioned is capable of acting at a distance?

10. How would you prove that heat and electricity give rise to force? State points of agreement and difference between electrical force and gravity.

11. State fully the points of difference between the *mass* and the *weight* of a body.

12. An ivory ball is let fall from a height of 3 or 4 feet upon a marble slab which has its surface covered with lampblack. After rebounding, a large black patch is seen upon the ball. Next, the ball is allowed to fall from a greater height upon the same slab, and the patch is seen to be larger than before. What do these two experiments prove? (Try them if possible.)

CHAPTER II

WORK DONE BY FORCES—FRICTION

Experiment 1.—Hold with one hand a wooden scale 2 or 3 feet long perpendicular to the table and with one end touching it, and lift with the other hand a mass of 1 lb. from the table through exactly 1 foot as measured by the scale.

The body before us is acted upon by gravity whether it is at rest upon the table or held by our hand. In one case gravity is *balanced* by the resistance of the table, in the other case by the force applied by our hand; and it is easily seen that whenever a body is prevented from falling vertically downwards by some other force acting vertically upwards, this force must be equal to the force of gravity acting upon the body, that is, to its weight. Thus we are enabled to compare different forces with gravity and hence measure their magnitude. In practice we always compare different masses at the same place by "weighing" them, that is, by comparing the actions upon them due to gravity, denoted by their "weights" at that place. Hence it has become customary loosely to speak of a mass or quantity of matter as a weight; it is thus that a mass of 1 lb. is called a "weight of 1 lb.," double that mass a "weight of 2 lbs.," and so on. This is of course incorrect. But the expression "weight of 1 lb.," besides being incorrectly applied to a certain definite mass called 1 lb., is also used to denote the force with which the earth attracts this mass at the place in which we are; and other forces are measured in terms of this as a unit. In strictness this ought never to be done, for the reason already explained in the preceding chapter. The *weight* of the *mass* of 1 lb. depends upon the place at which the 1 lb. happens to be, and speaking of a force as the weight of a certain number of pounds, or ounces, or grains, is just like speaking of a distance as a certain number of hours' or minutes' walk. This is often not only very convenient for practical purposes, but it makes you realise better what you are measuring; nor is it ever very indefinite, because, as a matter of fact, the weight of a pound, like the pace at which a man walks, does not vary enormously; but it is essentially inaccurate, and therefore out of the question whenever accuracy is of importance.

At present, however, for the sake of making our statements more easy to follow, we will say that the force required to hold this mass in our hands is equal to the weight of 1 lb. Now when we lift the body from the table to a height of 1 foot above it, as we have done, we have not only balanced the force of gravity, but have moved the body through 1 foot against the continuous action of gravity by *moving our hand* with the body in it vertically upwards. In this case we say that *work has been done* by the force which we have applied for lifting the body, and this work has been done against gravity. And generally, *whenever a force causes motion, work is done*.

We shall gradually become acquainted with a great variety of modes in which work is done by the different forces of nature. The most simple kind of work is that of lifting a body vertically upwards against gravity, and the work done in this manner serves to measure the amount of work done in all other cases, which is a matter of the greatest practical importance. In our case, having raised a weight of 1 lb. through a space of 1 foot against gravity, we say that we have done *one foot-pound of work.*

Experiment 2.—Lift the same mass from the table through 2 feet, and afterwards through 3 feet.

The body to which our force is applied has now been moved first through twice, and then through three times the space through which it was raised in the preceding experiment; hence the work done against gravity must be first twice, and then three times that done previously; and since raising this mass through 1 foot requires 1 foot-pound of work, it follows that we have done in this experiment first 2, and then 3 foot-pounds of work.

Experiment 3.—Lift a mass of 2 lbs. through exactly 1 foot, as before.

This must clearly be the same as lifting two single pounds each through 1 foot; we have therefore in this case done 2 foot-pounds of work exactly as in the preceding experiment, when we lifted 1 lb. through 2 feet. Similarly when we lift 3 lbs. through 1 foot, the work done is 3 foot-pounds.

From this it follows that when force is applied to move a body against the force of gravity, the work done is found by multiplying the force of gravity which resists the motion, that is, the weight of the body, by the space through which it has moved.

Experiment 4.—Place a square block of wood, which has one side rough and another planed, with the rough side upon a rough horizontal board, about 3 or 4 feet long and as wide as the block, and push it with your hand gently from one end of the board to the other.

Notice that force must be applied to push the block along; if your hand, which applies the requisite force, is removed, the block will stop and remain at rest. In this case there is no need for force to be applied by you to support the body against gravity, for the weight of the block is supported by the solid board upon which it is placed. Hence we conclude that the force necessary for moving the body along is required to overcome some resistance which does not arise from the action of gravity.

Experiment 5.—Repeat Experiment 4 after placing upon the block a very heavy mass, say 15 or 20 lbs.

You will find that a considerable effort is now required to push the block along, much greater than that needed in the preceding experiment. In this experiment also the action of gravity, though much greater than that upon the smaller mass used in Experiment 4, is balanced by the resistance of the board; but the force which resists our motion in both experiments has this in common with gravity in its action, that the force increases when the mass increases upon which it acts.

Experiment 6.—Place the block with its planed side upon a planed or polished board, and repeat Experiments 4 and 5 in succession.

The resistance now felt is in both cases, whether we move the block by itself only, or when heavily weighted, much less than that which is encountered in Experiments 4 and 5. Hence we conclude that the action of the force with which we have to deal depends upon the state of the surface of either body; if the two surfaces are rough the force acts very strongly and offers considerable resistance to any attempt to move one body upon the other; but if the two surfaces are very smooth the resistance to be overcome is very much diminished. Yet even now we cannot fail to observe that the force is very perceptibly larger when a larger mass is moved.

Further, when the block only is moved along, and the hand suddenly withdrawn, the motion will not stop at once, as in Experiment 4, but the block will go on moving through some distance; this will also happen with the weighted block, but to a less perceptible degree when the motion is slow, more so when the hand is pushing it along somewhat quickly. These last observations will be referred to in the next chapter.

Experiment 7.—Place the block with its smooth side upon one end of the planed board. Raise that end gently with the hand until the block begins to slide down the board. (Fig. 1, I).

As long as the board is in a horizontal position its resistance is capable of balancing the force of gravity, for this force acts vertically downwards, while the resistance of the board acts vertically upwards. But when the board is

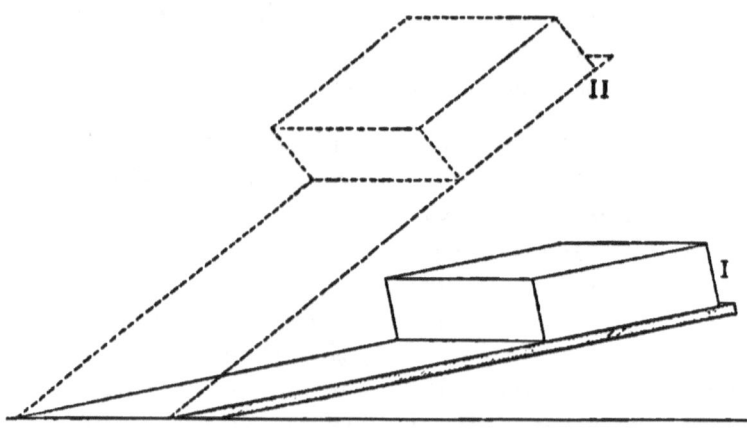

Fig. 1.

inclined ever so little by being raised at one end only, as in our experiment, its resistance no longer acts in the same direction as the force of gravity, and is therefore no longer capable of supporting the whole weight of the block; hence the block obeys the action of gravity and slides down. Now this should happen at once, as soon as the board takes up an inclined position. But our experiment shows that we may raise the end where the block is several inches before motion begins. Hence some force acts upon the block which prevents the block from sliding, and which helps the resistance of the board in overcoming the action of gravity. This force is called *friction*. It invariably acts, though to a less or greater degree depending on the kind of surfaces and their state as regards roughness or smoothness, when one of two bodies which are in contact moves with respect to the other, as we have seen in Experiments 4, 5, and 6. But it also prevents motion altogether in certain cases, as in the last experiment before the block began to slide. Thus friction is an example of a force which checks motion, or alters the rate of motion; but it is not capable of producing motion. In moving the block, weighted or unweighted, we are doing work against the resistance of friction, as we were doing work against gravity when raising a mass.

Experiment 8.—Repeat Experiment 7, but place the rough surfaces of the board and the block in contact.

Observe that now the block will not begin to slide until you raise the board much higher than before, Fig. 1, II. The friction between rough surfaces is greater than that between smooth ones. Yet it is easy to see that when the block is sliding the friction is less than gravity. For if the friction were equal to, or greater than the weight of the block, it would clearly be possible to raise the board until it is perpendicular, the block being wholly supported against gravity by friction; but this is not the case.

Experiment 9.—Repeat Experiments 7 and 8, using the weighted block in each case.

In either case it will not be possible to raise the end of the board higher than in Experiments 7 and 8, when the block was unweighted, to make it slide. We have already seen in Experiment 5, that the friction between the surfaces became greater when the block was weighted, that is, when its pressure upon the board was increased. From this experiment we learn that when the pressure is increased the sliding between the same two surfaces begins at the same angle of inclination as when the pressure is less. It follows from this that when the pressure increases the friction increases in the same proportion, and we have thus proved by our experiment the very important law,—that the friction between two surfaces is proportional to the pressure between them.

Experiment 10.—Examine with a magnifying glass the smooth surfaces of the block and the board.

You will find them by no means so very smooth as they appear to the unaided eye. Little inequalities will appear everywhere just like those seen with the naked eye upon the rough surfaces of both bodies, only smaller. From this we conclude that in moving one body upon another the rough projecting parts of one surface catch against the rough parts of the other and hinder it in moving along. But even if it were possible to produce perfectly smooth surfaces, we should thereby only increase those forces which act between matter in contact, called *cohesion* and *adhesion*, the action of which has already been described in Part I. These forces also check motion, and they increase in their action, the more polished and smooth and therefore close to one another two surfaces are. Hence we would still have some hindrance to motion of the nature of friction.

QUESTIONS ON CHAPTER II

1. Mention cases in which a force is doing *work*. Is work being done when you press very hard against the wall of a house?

2. How much work is done when 10 lbs. are raised 10 feet? How much when 100 tons are lifted 1 foot high?

3. A man whose weight is 170 lbs. goes up to the top of a mountain, which is 2800 feet high. Is work done in this case? How much?

4. In the last example, what force is doing work? Against what force is work done?

5. Why is it more difficult to move a body upon a rough surface than upon a smooth one?

6. Mention bodies which you would expect to move easily upon one another; mention others which it would be difficult to set in motion upon one another.

7. A large pane of plate-glass is placed upon the table, and another similar pane upon it. Do you expect to find it easy or difficult to push the upper pane with the fingers so as to make it slide upon the lower one? Try the experiment, and explain the result.

8. State the principal law of friction, and explain how it is proved by experiment.

9. What is the cause of friction? Why do we expect that the operation of greasing or oiling will diminish friction between bodies?

10. How do we prove that the force of friction between the two bodies in Experiment 8 must be less than the weight of the moving body?

11. A carriage is drawn along a level road, and the friction is $\frac{1}{40}$ of the weight. If the weight of the carriage is just 1 ton, how much work is done by the horses during a drive of 5 miles?

12. What use is made of friction on railways? Give other examples of the use of friction.

CHAPTER III

INERTIA—WORK—ENERGY

Experiment 1.—Press the point of a tack gently into a block of wood so that it just keeps upright by itself. Raise a mass of 1 lb. exactly 1 foot above the head of the tack and let it fall straight upon it.

Here a mass is set in motion by the force of gravity, hence work is done. But against what kind of resistance is work done in this case? In order to answer this question we must first of all consider that, so far as our experience teaches us, no mass ever begins to move of itself; whenever a mass at rest is set in motion some force has been acting upon it. Next, all our experiences of surrounding objects and the changes which they undergo as regards their state of rest or of motion invariably show that the larger a mass is the larger is the force required to move it when at rest and to stop it when in motion. In our previous experiments a mass of 2 lbs. maintained in our hands against gravity required a larger effort on our part than a mass of 1 lb.; a mass of 3 lbs. a larger effort than a mass of 2 lbs., and so on. It follows that gravity must act more strongly upon 3 lbs. to set it in motion and keep it moving than upon 2 lbs., and upon 2 lbs. more strongly than upon 1 lb. Hence gravity is doing work in our experiment against the resistance which our mass offers to the change of its state of rest into a state of motion; for if we suppose that at the instant when our hand releases the mass gravity altogether ceases to act upon it, there will be no reason why it should commence moving.

This resistance of matter to any change of its state of rest or motion is called the *inertia* of matter, and the quantity of matter in a body, or its "mass," is measured by the amount of this resistance to a change of given magnitude. Thus, for example, a mass of 2 lbs. possesses twice the inertia of a mass of 1 lb., or offers to every force in nature, of whatever kind, twice the resistance which is offered by a mass of 1 lb. to any change of state as regards rest or motion.

Our experiment, moreover, teaches another equally important principle. When our falling mass reaches the head of the tack it drives the point a certain distance into the wood. Here work is done against the force of cohesion, which, as we know, acts between the particles of the solid wood. In order to separate these particles another force must come into play, and just as we asked before against what force work is done by gravity when a body falls, so we must ask now what force is doing work against cohesion? The separation of the particles of wood by the tack when it is driven in cannot be due to gravity; for if we simply place our mass of 1 lb. upon the head of the tack and allow it to remain there for any length of time it will undoubtedly exert pressure in consequence of the action of gravity; but this pressure will not be sufficient to drive the tack into the wood to an appreciable extent. The work which is done in our experiment against cohesion is obtained from our mass solely because it is *in motion* when it strikes the head of the tack.

We shall see in the remainder of this book numerous instances of matter in motion being capable of performing various kinds of work, and we have already met with a few examples. Thus in Experiment 4, of Chapter I., the marble falling upon the table rebounded and rose again; here elasticity was doing work against that very force, gravity, which caused the marble to fall; but the elasticity was called into play by the compression of the particles of the ball due to the fall, that is, to its being in motion when it reached the table. If the ball is lowered gently till it touches the table it will neither be sensibly compressed nor rebound and rise again.

The capacity for doing work is called *energy*; hence a body capable of doing work is said to possess energy. In Experiment 6, of the last chapter, we pushed a wooden block along a polished surface, doing work against friction; when our hand was taken off, the block possessed energy in virtue of its being in motion, and it went on by itself, going over a short space and doing more work against friction. When afterwards the rough surfaces of the block and the board were in contact, the friction to be overcome was greater, hence the energy possessed by the moving block when left to itself was insufficient to carry it much farther, and it appeared to stop at once, though in reality it went on over some little distance, so small as to be insensible.

Experiment 2.—**Repeat the preceding experiment but use two tacks, a few inches apart; drop upon one a mass of 1 lb. from a height of 2 feet, and upon the other a mass of 2 lbs. from a height of 1 foot.**

The work done by gravity upon both masses is the same, viz., two foot-pounds, and we shall find, first, that both tacks are driven into the wood as nearly as possible to the same depth, and secondly, that in both cases the tacks have passed deeper into the wood than the tack used in the preceding experiment. Now, upon both masses used in the present experiment the same work was done; they also manifestly possessed the same energy, as shown by the equal depths to which the tacks were driven into the wood; hence we may conclude from this and the last experiment that the energy possessed by a moving body depends upon the work done upon it; or in other words, the more work is done upon a body the greater becomes its capacity for doing work itself; thus work and energy are measured in the same way. Hence a stone which has a mass of 10 lbs., and drops from a height of 100 feet, will be capable of doing 1000 times as much work when it reaches the ground, as a mass of 1 lb. dropping from a height of 1 foot. We may say of the first body that it has an energy of 1000 foot-pounds; of the second body that its energy is 1 foot-pound.

The question will naturally arise here, does our mass of 1 lb., when it has fallen through 1 foot, really do 1 foot-pound of work? The answer is, yes. No doubt it would be difficult to prove in this particular experiment that the work done when the tack, in entering the wood, forces its way through it by overcoming cohesion is indeed 1 foot-pound. Still we shall have other opportunities of proving that *energy is indestructible*.

But one point must not be overlooked at the stage at which we have arrived. When the mass was 1 foot above the head of the tack, it was lifted to that position by our own effort; we did 1 foot-pound of work before the mass was placed in a favourable position for doing work by falling down again, and thus we have reason to consider that the tack was finally driven into the wood by the work originally done by ourselves in raising the mass, its subsequent fall being only an intermediate link in the chain of events. A mass upon which 1 foot-pound has been thus expended in order to lift it against gravity, though not actually in motion, must therefore be considered as possessing so much energy. This kind of energy is called *energy of position*, while that kind of energy which is due to actual motion may be called *energy of motion*.

Experiment 3.—Boil a little water in a test-tube of stout glass, or copper, provided with a piston (Fig. 2).

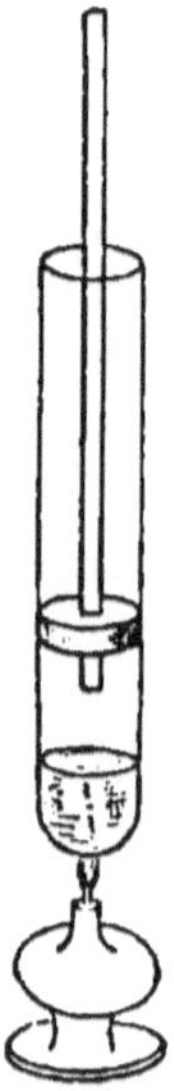

Very soon after the water has commenced to boil the piston will begin to move upwards, and will fly out of the tube if the boiling is continued. Here work is done against gravity, for the piston is raised; and, since there is friction between it and the sides of the tube, work is done in overcoming this friction; but, in addition, work is done against the pressure of the atmosphere; for the air above the piston is lifted and pushed aside while the piston moves upwards. There is thus a considerable amount of work done in this experiment, and the question arises how it is accomplished, that is, what energy is available for doing it. At first sight we might be disposed, from our experience of the expansion of air by heat, to answer: the air between the piston and water is heated, and expands; in other words, the particles of the heated air in tending to occupy a larger space than before, push the piston upwards. This explanation would, however, be insufficient to account for the result, as can be shown by a further experiment.

FIG. 2.

Experiment 4.—Repeat the previous experiment without any water in the tube.

If the piston fits pretty closely in the tube it will probably not move at all; or if it was well greased before inserting it, so as to render it capable of moving easily up and down with a little pressure, it will only move through a very short distance and then stop. Hence we conclude that the expansion of the air below the piston is not sufficient to account for the work which, as we have seen, was done in the last experiment. Indeed, as we shall have an opportunity of proving in Part III. of this work, the increase in volume due to heating the quantity of air beneath our piston would be insufficient to lift it to the top of our vessel, even if there were not the least friction between the piston and the sides of the tube.

In order to explain what we have seen, we must recollect, from previous experiments, that water when heated is converted into steam. In those cases in which we have boiled water, the steam was always observed to be in motion; it either left the vessel at the mouth and mixed freely with the surrounding air, or it passed into another vessel, whither it was led by means of tubing, for some definite purpose. Now considering that as long as there is water to boil, more and more steam is produced, and that this steam when once formed is a gas which expands when it is further heated, we must look upon the space between the water and the piston as filled with a comparatively enormous volume of gas contained within a very small closed vessel; such a bulk of gas will possess great energy of position because a great deal of work must have been done previously in some way or another to compress it into a small space, and when capable of moving a great deal of work will be done by it.

This is the case here. The steam up to a particular instant possesses energy of position, and this is employed to give to the piston energy of motion, and to do work upon it by pushing it forwards against gravity, friction, and the atmospheric pressure; if the piston were fixed, no doubt the vessel would be shattered to pieces as soon as the energy of the steam is capable of doing the work required for the purpose. But whence do the steam and the enclosed air derive this energy? The answer is obvious: this energy is communicated as heat, which in its turn is derived from the chemical union of the spirits of wine used in our experiment, which is composed of the same elements as a candle or wood, with the oxygen of the air. It is heat alone which gives motion to air in the process of expansion and to water in the formation of steam, and endows both substances with the capacity for doing work in some form or another, that is with energy.

Experiment 5.—Repeat Experiment 3, but as soon as the piston has moved nearly up to the mouth of the tube plunge the whole into a tumbler of cold water.

The piston will at once begin to move downwards again. The steam and the air are cooled by the surrounding water, the steam is therefore condensed and returns to the liquid state while the air contracts to its original volume. As a consequence the pressure of the air outside pushes the piston down to its former position.

This experiment is calculated to prove very strikingly the fact that in our previous experiment work was done against the pressure of the atmosphere. This pressure is due to the action of gravity, and when work is done against it, as by our piston when it rose, the air is placed in the condition of possessing energy of position, just as the mass raised in former experiments; as soon as the steam and the air returned to their original state, the energy of position became energy of motion, as in the case of the mass allowed to drop when deprived of its support, and work is now done against friction.

We know that when a solid body is heated it expands, though less than a liquid or a gas. Now such a body must also be doing work against gravity in pushing aside and partly lifting the air around itself, while it expands. But, further, though the body as a whole may remain at rest, the expansion can only consist in a general movement of those smallest particles of which we suppose all bodies to be ultimately made up, and which are called *molecules* (from the Latin *moles*, a mass). It is solely in this vigorous motion of its molecules that a hot body differs from a cold one, and as this molecular motion is capable of doing work we are justified in saying that *heat is a form of energy.*

Experiment 6.—Place a piece of lead, measuring about an inch each way, upon an anvil or a flat stone, and hammer it steadily and vigorously for a minute or two.

The piece of lead will become sensibly heated. Here the hammer first of all possesses energy of motion, but loses it, and work is done upon the lead. Only a small part of this work is spent in altering the shape of the piece of lead; the greater part of it has been employed in heating it, that is in giving motion to the molecules of the lead. We know already that heat is a form of energy, and that it is capable of doing *mechanical* work, as in causing steam and air to push a piston through some distance; now we see that the energy of the moving hammer which is generally in practice converted into mechanical work, as in driving a nail into a board, here is partly converted into mechanical work, partly into another form of energy, namely heat.

Thus we learn, and we shall see more clearly farther on by a great many such changes, that energy in nature assumes a variety of forms, and that many of these are capable of being converted into one another.

QUESTIONS ON CHAPTER III

1. What is meant by *inertia*? How would you explain the following facts derived from experience :—
   - (a) It is an advantage to run before a leap.
   - (b) It is safest to skate quickly over thin ice.
   - (c) When a horse bolts, or refuses to jump, the rider is frequently thrown. In which direction will he be thrown in each of these two cases?

2. Supposing that you raise a mass 2 feet above the table and allow it to drop, what would happen if exactly at that instant gravity ceased to act? What would happen if gravity ceased to act after the mass has fallen down through 1 foot?

3. Describe in your own words the object of Experiment 2, and the conclusions drawn from it. What is meant by *energy*?

4. Describe clearly the differences between *force*, *work*, and *energy*. Give examples of a correct use of these terms.

5. Explain why a boiler full of air is not likely to burst when heated, but will explode if it contains some water, unless it has an outlet for the steam.

6. How can we prove that a heated body is doing work against gravity?

7. Mention the different bodies, possessing energy, with which you have so far become acquainted.

8. How is it proved that the energy of a moving body may be converted into heat?

9. Is the whole energy of the hammer in Experiment 6 accounted for by the flattening and heating of the piece of lead? If not, what becomes of the remainder?

CHAPTER IV

CENTRE OF GRAVITY, OR CENTRE OF MASS—EQUILIBRIUM

Experiment 1.—Punch a small hole close to one corner of a square or rectangular sheet of stout pasteboard, or tin-plate, ABCD, Fig. 3, and suspend it by a thread from a suitable support, so that it may swing quite freely.

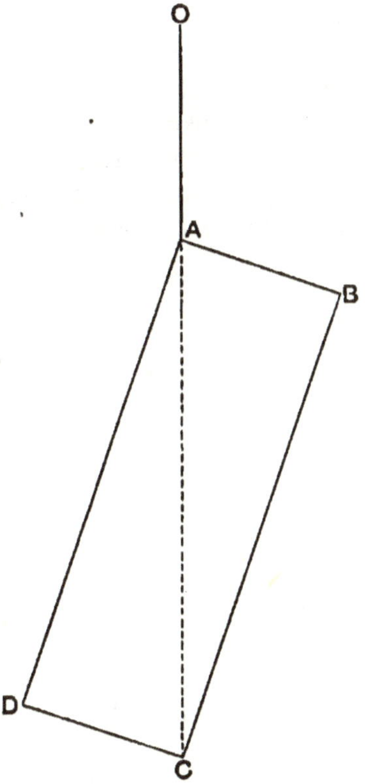

When the suspended mass is at rest two forces are acting on it, gravity and the resistance of the support which is transmitted by the thread; gravity acts vertically downwards, and as there is no motion the resistance of the support must act vertically upwards. A mass thus acted on by forces which balance one another so as to produce rest is said to be in *equilibrium*.

In the position of equilibrium, the figure being suspended at A, the opposite corner C will be found to be vertically beneath the point of suspension A. Now we know that gravity is acting upon each portion of our

FIG. 3.

mass, for if we were to cut off any portion of the suspended body, that portion would simply fall; therefore the resistance of the support transmitted along the thread clearly balances the action of gravity exerted upon the different particles of the mass taken together. Hence the action of gravity is the same as if the whole mass were collected at some point in the line from A to the opposite corner C, which, as we see, divides the figure geometrically into two equal parts. We say in such a case that the *resultant* of the forces due to gravity acts along the line AC, denoting by the term resultant a single force which produces the same effect as any number of forces acting upon the same mass are producing jointly.

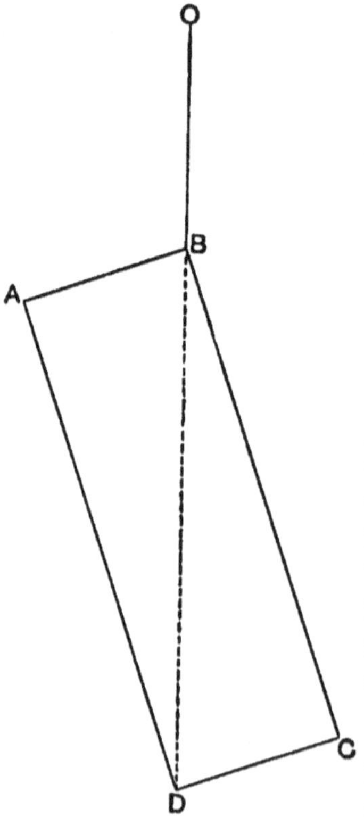

Experiment 2.—Displace the suspended mass by taking hold of it at any point and lifting it a few inches; then leave it to itself.

The body will after each displacement return to the position assumed at first. Equilibrium of this kind, in which a body rests steadily and requires force to move it, and when it is moved returns or tends to return to its former position, is called *stable equilibrium*.

Experiment 3.—Fasten the thread to the body at the corner B and suspend it as before (Fig. 4).

Fig. 4.

The opposite point D will now be found vertically

beneath B. Thus the resultant force due to gravity now has the direction of the line BD which crosses AC.

If the body is displaced it will return to its original position, thus proving that this position is again one of stable equilibrium. Thus a body may be in stable equilibrium in different positions.

Experiment 4.—Draw straight lines on the sheet, joining AC and BD. At the point G, where these lines cut one another, make a small hole, fasten the thread at that point, and suspend the body again (Fig. 5).

Our mass will now be again in equilibrium, but when

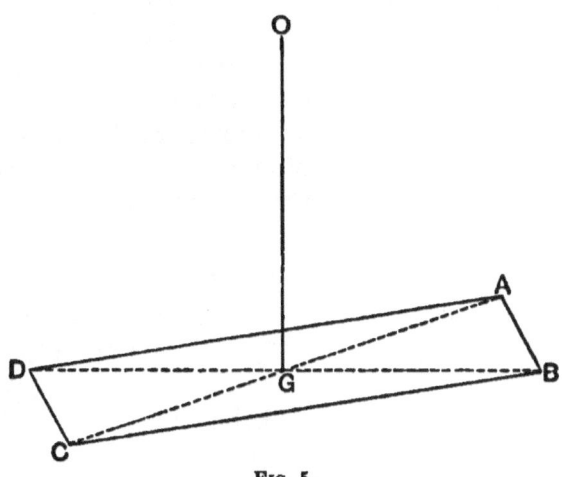

Fig. 5.

displaced will remain in any position we assign to it, in other words, it will rest in any position indifferently, and will require very little force to move it from one position to another. This kind of equilibrium is called *neutral* or *indifferent equilibrium.*

It is clear from what we have just seen that the point G, at which our body is now suspended, must be so situated with regard to all other portions of matter in the body that the whole action of gravity upon the mass is counteracted if this one particular point is supported, that is, the whole weight of the body may be considered as if collected in this one point. The point which has this property, and every mass, no matter what be its shape, has such a point, is called the *centre of mass*, or *centre of gravity* of a body.

Further, by considering the position which the point G had in Experiments 1, 2, and 3, we find that when the body is suspended *above* its centre of gravity the equilibrium is stable; and from the present experiment we learn that when the body is supported *at* its centre of gravity the equilibrium is neutral.

Experiment 5. — Punch a hole at some point F, between G and one of the corners, tie a loop to each end of a long piece of thread, draw the thread through F, and suspend the figure by both loops (Fig. 6).

If carefully suspended the body will be at rest in this position also, but the equilibrium will be far from stable: the slightest disturbance causes the figure to swing round, and to assume a position of stable equilibrium in which the centre of gravity is below F. This kind of equilibrium, in which the body is easily moved, and if moved a little way tends to move still farther from its original position until it takes up a new position, viz. that of stable equilibrium, is called *unstable equilibrium*.

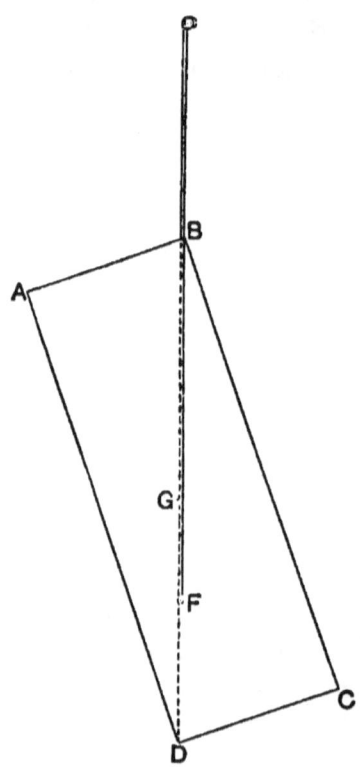

Fig. 6.

CHAP. IV   CENTRE OF GRAVITY—EQUILIBRIUM   31

Experiment 6.—Cut out a body of the shape shown in Fig. 7, and suspend it in the two positions, A and B, at the point S.

The centre of gravity of a regular body like this will be at the centre G, because the mass is equally distributed around it. Observe that the body will be in equilibrium in neither position. It is neither in stable nor unstable equilibrium, because the vertical line through G does not pass through the point of suspension as in all cases considered in the preceding experiments; and it is not in neutral equilibrium because it is not suspended *at* the centre of gravity. The body not being in equilibrium

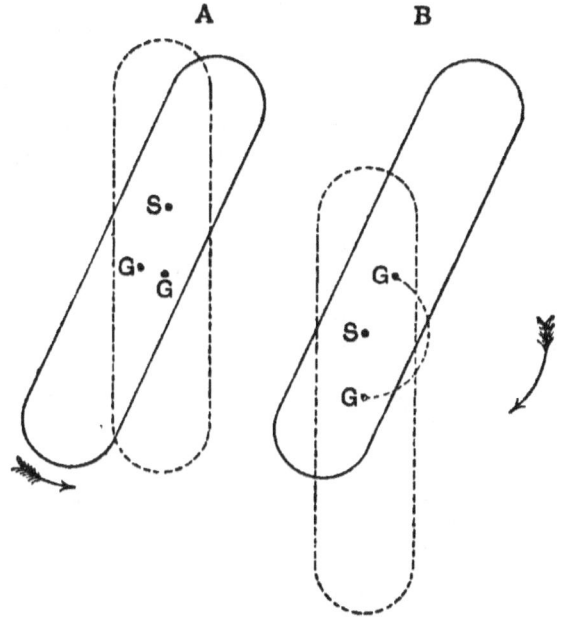

FIG. 7.

of any kind, is acted on by gravity until it assumes the position of stable equilibrium. Now the space which the centre of gravity must describe before stable equilibrium is attained is very unequal in the two cases A and B, as can be seen in the experiment, and as is shown by dotted lines in the figure. In both positions gravity tends to turn the body in the direction indicated by the arrow, until the centre of gravity G falls on the vertical line through the point of suspension S. In the position A this is soon accomplished, but in the position B it takes an appreciable time, and it is for this reason that such bodies as long poles, sticks, or similar objects may be "balanced," on the tip of the finger, for instance. Such bodies are in unstable equilibrium, being supported by a small surface below their centre of gravity; hence the slightest disturbance makes them fall over, but, the centre of gravity being a good way above the point of support, the time required for returning to the position of stable equilibrium is comparatively great, and so is sufficient for bringing the point of support below the centre of gravity again; thus the body can be maintained in its unstable position for any length of time.

Experiment 7. — **Place the wooden block used in Chapter II. with one of its narrow faces upon the table, and turn it slowly with your hand about one of its edges, as E, until it falls (Fig. 8, A, B, C).**

We may assume that G is the centre of mass, and that the resultant of gravity acting along GA is counteracted by the upward pressure of the table at A. We should thus have a case of unstable equilibrium, as the body is supported beneath its centre of mass. Yet our experiment proves that a body like this, resting on an extended surface, behaves when turned about one of its edges exactly like a body in stable equilibrium, and will continue to do so up to a certain point, viz. while the line GA does not pass beyond the edge E. In the case B the position of the body will correspond to the condition of stable equilibrium: the body when left to itself returns to its former position.

In the case C, when the vertical line GA has moved to the right of E, the body will turn over, that is, it will take up a new position of rest, indicated by dotted lines. This new position, however, is simply some other position like

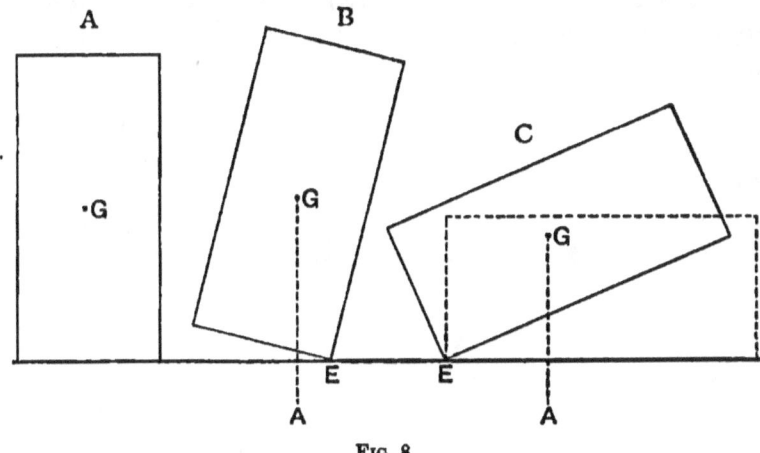

Fig. 8.

the original one; not being, as in the case of a body previously in unstable equilibrium, the reverse of the previous one, viz. that of stable equilibrium.

Observe, however, that in moving the body from position A to B, and from B to that position when the vertical line GA just passes the edge E, the centre of gravity must be raised, that is, work must be done against gravity. Now the work to be done in raising the centre of gravity of a mass depends clearly first of all on the total action of gravity upon it, that is, upon its weight. If our block were iron or lead, more work would have to be done to move the body through the stage B, until it assumes the new position in which it begins to be upset. Hence, the greater the work required for upsetting a body, the greater is its *stability*, and we see that other things being the same, the heavier a body the greater its stability.

The work required depends however not on the weight

alone, but also on the vertical space through which the weight has to be moved.

Experiment 8.— Place a wedge-shaped body, like that in Fig. 9, and in the same position upon the table, and upset it twice in succession, first by moving its upper edge gradually to the left, then by moving it to the right.

Observe that more work is required for upsetting the body on the right side than on the left, the reason of this will easily appear from a glance at our figure. The centre of gravity, G, of the body is nearer to the left edge than to the right, and in turning the body on that side the work

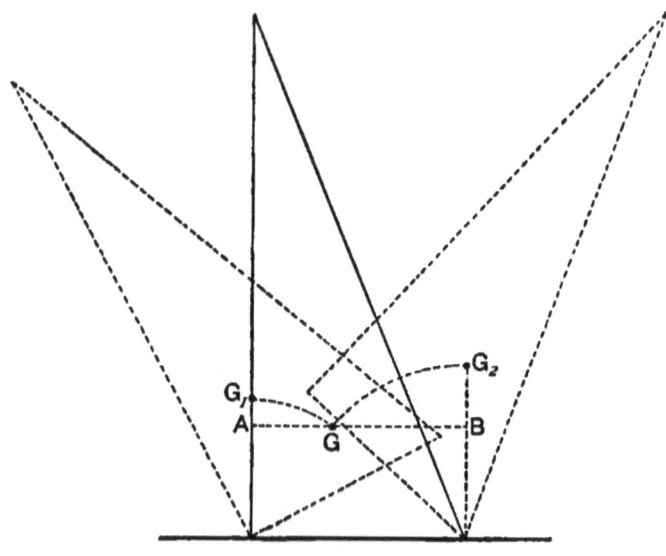

Fig. 9.

done consists in moving the centre of gravity through the arc GG_1, or in raising its weight vertically through the space AG_1; but if overturned about the other edge, the centre of gravity will have to describe the arc GG_2, and the

CHAP. IV CENTRE OF GRAVITY—EQUILIBRIUM 35

weight is raised through the larger space BG_2. It follows that the stability of a body is increased if the centre of gravity is in all directions as far as possible from the edges round which the body can turn, that is, if the base on which the body rests is as extended as possible.

Experiment 9.—**Place a body, half of wood and half of iron or lead, first with the metal then with the wood uppermost (Fig. 10, I, II), and turn it about on edge until it is upset.**

It will be found at once that the stability of the body is

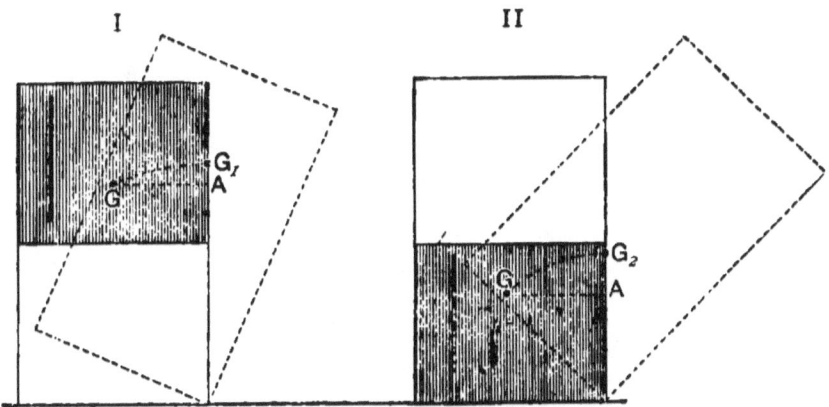

Fig. 10.

far less when the heavier portion is uppermost, than when it is below the lighter part. The centre of gravity of a body like this will not be in the centre, but will fall within the mass of iron or lead, at G. In the position I the centre of gravity is higher above the base than in the position II. In the first position, if the body be turned about the edge, the centre of gravity describes the arc GG_1, and the weight is raised through the space AG_1. In the second position the weight is lifted through the space AG_2, and it can easily be shown by geometry that the line AG_1

is shorter than the line AG_2. It follows that the lower the centre of gravity is situated in bodies which are equal in other respects, the greater is the work which must be done upon the body in order to upset it, that is, the greater is its stability.

QUESTIONS ON CHAPTER IV

1. Describe cases, taken from your own experience, in which a body is in *equilibrium*, while acted on by several forces.

2. What is meant by the *resultant* of forces ? Two players push a ball in opposite directions, one applying exactly double the force of the other ; what is the resultant in this case ? What would it be if they pushed both in the same direction ?

3. Describe how we may find by experiment the centre of gravity of an irregular piece of sheet-iron or cardboard. How would you prove that the point you found was the true centre of gravity ?

4. When is the equilibrium of a body *stable* ? When is it *unstable*, and when *neutral* ? Give illustrations of each kind of equilibrium.

5. Explain why it is possible to balance a long pole on the tip of a finger.

6. What is meant by the *stability* of a mass ? Give examples of bodies possessing considerable stability, and explain how this has been obtained in each case.

7. What is meant when we call a body *top-heavy* ? Give an example of such a body.

8. Which is more easily upset—
   (a) A cart loaded with iron, or one loaded with the same weight of hay ?
   (b) A pillar of stone, or a pillar of wood of the same size ?

State the reasons for your answer.

9. Why is there so much objection raised to *deckloads* on ships ? What is the use of *ballast* in ships which have no freight ?

10. Why will a cask roll down a slope, while a flat box placed upon the same slope remains at rest?

11. Sketch and describe the position in which a loaf of sugar on the table would be (I) in stable; (II) in unstable; (III) in neutral equilibrium.

12. How would you place a book upon a table that it may be (I) in stable; (II) in unstable equilibrium? State your reasons.

CHAPTER V

THE LEVER

Experiment 1.—Place before you the apparatus of Fig. 11, see that the wooden bar AB is in a horizontal position, gently push down the end B with your finger, and let the bar come to rest again.

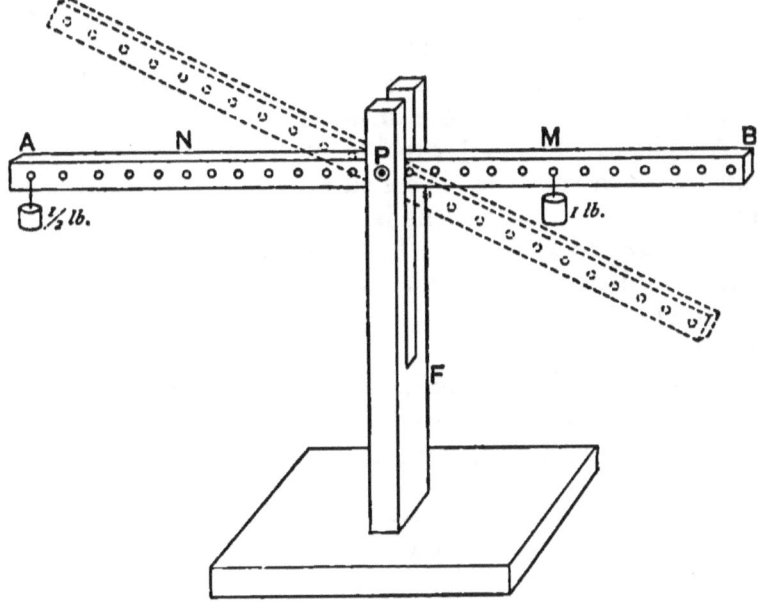

The bar AB is of uniform thickness throughout, and its

centre of gravity is just a little beneath the hole bored through it at a point midway between its ends. It is able to turn freely about a round smooth peg P, which passes through this hole and also through corresponding holes in the frame F. From what we have learnt in the preceding chapter it follows that the bar when in a horizontal position is in stable equilibrium, and will return to this position when either end is moved through any distance upwards or downwards and then let go again.

If the peg P passed exactly through the centre of gravity of the bar, the bar would be in neutral equilibrium and would not return to the horizontal position when displaced; for example, if depressed to the position indicated in Fig. 11 by the dotted outline of the bar, it would remain in that position. For our experiments this kind of equilibrium would be quite unsuitable.

A rigid bar like AB, capable of turning about a fixed point, as P in this case, is called a *lever*. The fixed point about which the lever turns is called its *fulcrum* (from the Latin *fulcire*, to support). If forces are applied on either side of a lever, the distances from the fulcrum to the points where these forces are applied are called the *arms* of the lever.

In the following experiments it is assumed that holes are bored through the lever at distances of 1, 2, 3, etc. inches, up to 12 inches, on either side of the fulcrum.

Experiment 2.—Suspend a mass of 1 lb. from the hole nearest to the extremity B, that is, 12 inches from the fulcrum.

The arm on the side of B sinks, the other arm rises. We are familiar with a very similar effect when placing a weight into one scalepan of a balance; that pan sinks, the other rises. The "beam" of an ordinary balance is indeed a lever like ours, with scalepans attached to either end. The reason that the weighted arm of our lever and the weighted scalepan go down is this: before any mass is placed upon one side, the centre of gravity of the whole is exactly beneath the fulcrum, but its position shifts in the direction of the mass as soon as this is suspended, and the beam moves as shown by the dotted part of Fig. 11, so that the new centre of gravity, which is then somewhere between P and B, may again occupy a position beneath the fulcrum, and thus bring the whole into stable equilibrium. The consequence of this is that the arm PB sinks.

When the arm PB with its suspended mass moves downwards, work is done upon it by gravity, and if we wish to bring back the lever to its horizontal position work must be done by applying force so as either to raise the arm PB, or to depress the arm PA. We may do this by our hand, that is, by applying muscular force at some point between P and B so as to lift the arm PB, or between P and A so as to push the arm PA downwards. By actually applying our hand for that purpose at different points between A and B, we shall at once perceive that the force required at different points is sensibly different: it is greater the nearer to the fulcrum our hand is applied, and less the nearer to the points A or B our muscular force is made to act.

Experiment 3.—Suspend successively different pairs of equal masses A and B on either side of the fulcrum.

Neither arm will now sink or rise, for there is no reason why the lever should move either upwards or downwards on one side or the other. Whatever work could be done on one side in consequence of the action of gravity upon the mass suspended at that side would also be done to the same amount on the other side, because the masses are equal; but if motion took place one mass would move upwards through the same space as that which the other mass moved through downwards. The consequence is that no motion takes place, and hence that equal masses at equal distances from the fulcrum are in equilibrium.

Nor can we doubt that what is proved by our experiment will be the case generally. A body will always be in equilibrium if the forces acting upon it are such that the work done by one or more of these forces, if motion takes place, is equal to that done by the other force or forces, provided that the motions, produced by these two sets of forces, would take place in opposite directions.

Experiment 4.—Suspend a mass of 1 lb. at the point B, and another equal mass at the point M, midway between the first mass and the fulcrum. Allow the mass at the end B to go down exactly through 1 foot, as measured by a scale, and then hold the end of the lever near B with your hand for a few minutes.

Observe now that while gravity has done one foot-pound of work upon the mass at B which is 12 inches distant from the fulcrum, it has only done half that work upon an equal mass at M, which is 6 inches from the fulcrum; for while the mass at B has moved downwards through one foot that at M has only moved downwards through one-half of a foot, as we may easily find by measuring with the scale the distance through which M has moved.

From this it follows that if we place at B one-half of the mass suspended at M, the same work will be done upon each mass through whatever distance the arm PB may move. Now we have seen in the preceding experiment that two equal masses, one at M 6 inches from the fulcrum, and the other at N on the other side of the fulcrum and at the same distance from it, keep the lever at rest. Hence we may conclude that if instead of the 1 lb. mass at N we place a $\frac{1}{2}$ lb. mass at A, 12 inches from the fulcrum, the lever will still remain horizontal, because the work done by gravity upon one arm, if any motion takes place, will be equal to that done by the other arm against gravity.

Experiment 5.—Suspend masses of $\frac{1}{2}$ lb. and 1 lb. at A and M respectively as in Fig. 11, and placing one of your fingers upon the end A push that end down a few inches, and then let go.

Very little force is required to push the end A down, sensibly less than was required previously before the $\frac{1}{2}$ lb. mass was suspended at A. Further, after swinging once or twice up and down, the lever will soon come to rest in a horizontal position.

We thus see that a small mass upon one arm of a lever may not only balance a larger mass upon the other arm, if the work done by gravity upon one is equal to that done by the other against gravity, but that a smaller mass may even be employed for lifting a larger one against gravity if the work done upon it is somewhat larger than that done by the greater mass whilst motion takes place. The slight push given to the end A of the lever in this experiment by muscular force may be made more constant and productive of more continuous motion by adding a small mass to our $\frac{1}{2}$ lb., and thus replacing muscular force by gravity.

Further, if we suppose our lever to have any length whatever, and its fulcrum to be at any other point, for example at the end A, if B be moved downwards or upwards through 1 foot, a 1 lb. mass placed at B will do or require 1 foot-pound of work; if placed at half the whole length of the lever from the fulcrum the same mass will only do or require one-half of a foot-pound, at one-third from the fulcrum only one-third of the same work, at one-fourth the length only one-fourth of the work, and so on. In other words, in *comparing* the work done upon different masses either balanced or set in motion by means of a lever we need not measure the actual distances through which the masses move or would move in case of motion, but we simply measure their distances from the fulcrum; the product of any one mass into its distance from the fulcrum may then be compared with the product of another mass into its distance from the fulcrum; if these products are equal and the action of gravity upon the two masses tends to turn the lever in opposite directions, we shall have rest or equilibrium; if these conditions are not fulfilled we shall produce motion.

Experiment 6.—Place 1 lb. at A, and find the points between P and B where masses of 2, 3, 4, 6 and 12 lbs. must be suspended to maintain the lever at rest. Then add each time at A the *smallest possible* mass which will make that end go down.

In the first experiment we have at A 1 lb., its distance from the fulcrum 12 inches; hence—
$$\text{product of mass into distance} = 1 \times 12 = 12.$$
On the other side of the fulcrum we have 2 lbs., and the distance of this mass from the fulcrum will be found to be 6 inches, when there is equilibrium; thus—
$$\text{product of mass into distance} = 2 \times 6 = 12.$$
Similarly in the succeeding experiments we shall find that—

3 lbs. must be suspended at a distance from the fulcrum of 4 inches
4 ,, ,, ,, ,, 3 ,,
6 ,, ,, ,, ,, 2 ,,
12 ,, ,, ,, ,, 1 ,,

for maintaining equilibrium, and we see that in each case the product of mass into distance on both sides of the fulcrum is equal to 12.

If care was taken to add each time as small a mass as will just suffice to set the lever in motion after obtaining equilibrium, we shall further find that this additional mass must be greater as we continue our experiments, that is, as we go on weighting the lever with larger masses. The reason of this is that the additional mass has not only to set the lever and the suspended masses in motion, but also to overcome the friction of the peg in its bearings, which constitute the fulcrum. When in our first arrangement we have a 1 lb. mass on one side and a 2 lbs. mass on the other, the pressure on the fulcrum is precisely 3 lbs., together with the weight of the lever, which is comparatively small; on the other hand, in our last arrangement we have over 13 lbs. suspended, hence the friction, which as we have learnt increases with the pressure, is appreciably greater than at first, and a somewhat larger mass must be added to overcome this friction and give motion to the whole.

The most important practical fact, however, which the preceding experiments teach us, is that we may gain a *mechanical advantage* by being able, with the help of the lever and other so-called machines, with some of which we shall soon become acquainted, to move a great mass and thus to overcome a large *Resistance*, by moving the point at which a small force, called the *Power*, is applied, through a proportionally larger space. In order to estimate the mechanical advantage of any given machine we have only to measure the spaces through which the Power and the Resistance are moved respectively in the same time, when the machine is working. The mechanical advantage is found by dividing the space through which the Power moves by the space through which the Resistance moves. Thus in our lever, when we have 1 lb. at A, and 12 lbs. at 1 inch from the fulcrum, the Power, viz. the 1-lb. mass, will be found to move through 1 foot, while the Resistance, viz. the 12 lbs. mass, moves through 1 inch. Hence—

$$\left.\begin{array}{l}\text{Mechanical}\\ \text{advantage}\end{array}\right\} = \frac{\text{Space through which the Power moves}}{\text{Space through which the Resistance moves}} = \frac{12 \text{ in.}}{1 \text{ in.}} = 12.$$

Fig. 12 exhibits a variety of forms which the lever assumes in its practical applications. On carefully studying these diagrams it will be seen at once that the fulcrum is by no means always *between* the Power and the Resistance. When this is the case the lever has two arms, and is said to be "a lever of the first order." The only other position which the fulcrum can have is one end of the lever. In that case the lever has only one arm and the Power is either at the other end, and the Resistance between Power and fulcrum, in which case it is a "lever of the second order"; or the Resistance is at the other end, and the Power between fulcrum and Resistance, which arrangement forms a "lever of the third order."

Levers of the first order are: a crowbar, applied to move or raise a heavy weight, as in Fig. 12, No. 14; and the handle of a pump (No. 9). Scissors (No. 4), pincers (No. 3), and other similar instruments, are "double" levers, being composed of two levers of the first order. A key (No. 6) is also a lever of the same kind, the Power being applied at both ends of the handle, the Resistance at the bolt which is pushed backwards or forwards by the wards. To the same kind belongs the lock itself (No. 7).

Levers of the second order are: an oar (No. 5), the fulcrum being at the end of the blade which is kept at rest by the pressure of the water, the hand of the oarsman exerts the Power, and the inertia and friction of the boat, which are overcome by pressure upon the rowlocks, form the Resistance. The chopping-knife (No. 12) has one end attached to the board as fulcrum, the Resistance of the substance to be cut being overcome by the Power applied to the handle. A very similar lever is No. 10, a piston worked up and down by a horizontal rod. In a wheelbarrow (No. 13) the fulcrum is the point at which the wheel presses upon the ground, and the resistance is the weight of the barrow and its load collected at the centre of gravity of the whole. Nutcrackers (No. 2) are "double" levers of the second order; the hinge which unites them being the fulcrum, while the Resistance is presented by the cohesion

CHAP. V THE LEVER 47

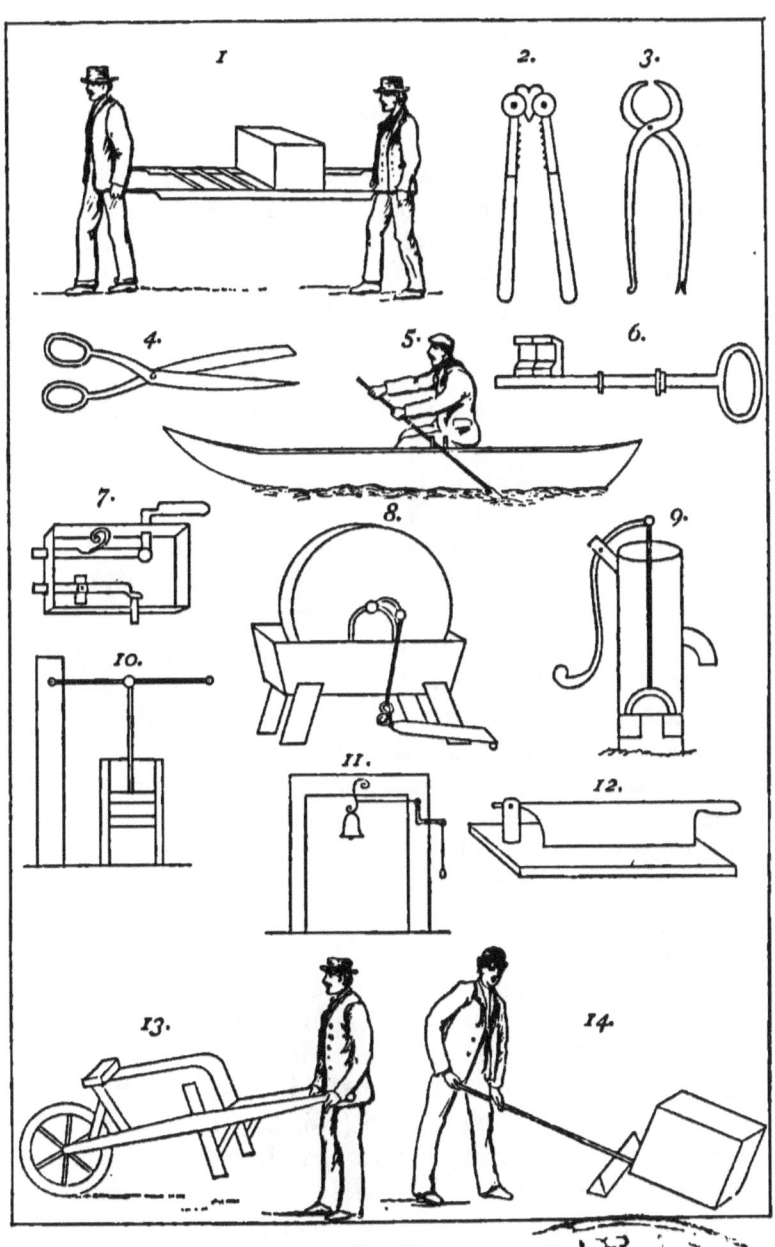

Fig. 12.

of the shell of the nut placed between them, and the hand applied to the end being the Power.

The hand-barrow (No. 1) may be considered as a lever of the second order. Each man acts as the Power in lifting the weight, and at the same time the hand of each serves as a fulcrum for the lever of which the hand of the other is the Power, and the weight of the barrow and load the Resistance. It is not difficult to see, from the facts established by our experiments, that if the load is fairly in the middle of the barrow, each man will bear just half the burden; but if the weight is placed nearer towards one end, then the man to whom it is nearest supports a greater load than the other.

The treadle of a grindstone (No. 8) is an example of a lever of the third order. The hinge at which it turns is the fulcrum, the foot applied to the treadle near the hinge is the Power, and the rod attached to the other end which turns the grindstone is the Resistance. Since the Power acts nearer to the fulcrum than the Resistance, it must be proportionally greater than it, and hence it acts upon the Resistance to a mechanical disadvantage. But we have seen from our experiments that the farther any point on a lever is from the fulcrum the larger is the space through which it passes when the lever turns, as compared with a point nearer to the fulcrum. It follows that in such a lever the speed with which the Resistance moves will be greater than that of the Power. Levers of the third order are therefore chiefly applied where the expenditure of a comparatively greater Power is of minor consideration, while the gain of speed and rapidity of motion for the Resistance is of more importance.

A bell-handle (No. 11), and the bolt of a lock (No. 7), also the handle of the pump (No. 9), are examples of "bent" levers, in which, besides gaining some mechanical advantage, the force which acts as Power produces an effect in a direction not originally its own. Thus a vertical pull at the bell-handle pulls a horizontal string or wire which rings the bell. In the lock the bent form of the lever enables us to move the bolt upwards by the pressure upon the handle applied downwards.

QUESTIONS ON CHAPTER V

1. A uniform bar of wood is suspended as in Experiment 1, but the peg passes through the bar below its centre of gravity. Why would such a bar be useless for our experiments?

2. Draw the outline of a balance having a weight in one scalepan only, and roughly indicate the position of the centre of gravity of the whole (I) before the weight was put into the scalepan; (II) after it was put in.

3. Describe the successive experiments we made for proving that a small mass upon one arm of a lever must be capable of balancing a larger mass upon the other arm. What condition must be fulfilled when this is done?

4. A straight uniform lever, 36 inches long, is supported in the middle. If a mass of 10 lbs. is placed upon one arm at a distance of 16 inches from the fulcrum, how much must be placed upon the other arm, at a distance of 10 inches from the fulcrum, if the lever is to remain in equilibrium?

5. Sketch as well as you can the separate appliances of Fig. 12, and denote in your diagrams accurately the positions of Fulcrum, Power, and Resistance.

6. Suppose our lever in Fig. 11 weighs 2 lbs. and is suspended by the peg passing through a hole 6 inches from the middle, instead of being supported at the middle. Why would the lever not remain horizontal in the new position? What would happen? How would you proceed to maintain it in a horizontal position if you had (I) a mass of 2 lbs. only; (II) one of 4 lbs. only, at your disposal?

7. What kind of levers are the following? State in each case the reason for your answer.

> (a) A spade when used to detach a portion of earth from the main mass by forcing back the handle.
> (b) A poker applied to raise fuel.
> (c) A door or gate moved on its hinges.
> (d) Shears used for shearing sheep.
> (e) A pair of tongs used for coals or sugar.

8. In a lever of the second order the distance of the power from the fulcrum is 30 inches, and that of the resistance from the fulcrum is 6 inches. Find the mechanical advantage.

CHAPTER VI

THE PULLEY

Experiment 1.—Suspend a pulley from a hook which is fixed to the cross-beam of a suitable frame. Attach one end of a flexible cord to a 1 lb. mass, pass the cord over the groove cut in the wheel and raise the mass with your hand, as in Fig. 13.

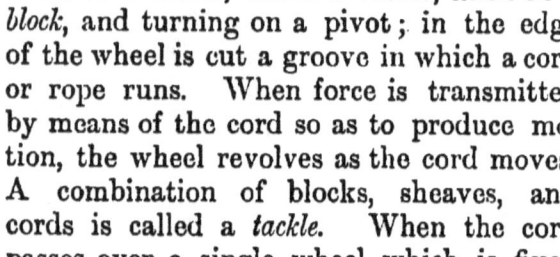

Fig. 13.

A pulley consists of a wheel, called a *sheave*, fixed in a *block*, and turning on a pivot; in the edge of the wheel is cut a groove in which a cord or rope runs. When force is transmitted by means of the cord so as to produce motion, the wheel revolves as the cord moves. A combination of blocks, sheaves, and cords is called a *tackle*. When the cord passes over a single wheel which is fixed in position, as in Fig. 13, the machine is called a *fixed pulley*.

Observe that by pulling the cord you are doing work because you raise the mass attached to the end of the cord against the action of gravity; if the mass is lifted through 1 foot, the end of the cord to which your hand is applied is also pulled through 1 foot; the force which must be applied by you is obviously as great as the weight, for this force is simply transmitted through the length of the cord, and if it were not equal to the weight the mass could not be lifted. Hence there is no mechanical advantage gained by this contrivance. Nevertheless, as our experiment proves, we are raising our mass by applying force in a direction more convenient to our position on the ground than if we had to raise the same mass by applying the necessary force without the help of the pulley. A fixed pulley, or a combination of fixed pulleys, as that shown in Fig. 14, is a machine which enables us to give the most advantageous direction to the force employed in certain cases for doing work, and hence possesses as much practical utility as one in which a smaller force enables us to balance or to move a great weight.

Fig. 14.

In Fig. 14 is shown an example of this application of fixed pulleys. The work done by a horse drawing the end of a rope in a horizontal direction is used for elevating a considerable mass to the top of a building, by means of the two fixed pulleys A and B suitably placed. In a similar manner very extensive use is made of fixed pulleys on ships, for spreading sails and hoisting flags on the yards and masts, by sailors pulling a rope on the deck.

We may here ask, what is the advantage of employing a pulley in such cases? Would not the more simple contrivance of a cord passing over a fixed peg answer the purpose of altering the direction of the applied force quite as well? The answer is: no doubt it would be possible to substitute a peg, or in fact any edge or solid angle, for changing the direction of the cord, and hence that of the applied force. But we must not overlook the amount of friction caused by pulling a cord over such a comparatively sharp body; even if the peg were rounded off smoothly, there would still be a great amount of chafing of the cord, especially if a heavy mass were moved repeatedly by means of it. Thus there is a great deal of work wasted in overcoming friction, and the rapid wear of the cord will soon cause it to break, however strong. The waste and danger are both reduced to a minimum by means of the pulley with its wheel; for the surface upon which the cord runs moves with it, so that the friction is the same as if the wheel were rolling along the cord, and this is comparatively very small. We shall in Part III. of this book have an opportunity of measuring some effects of friction and we shall then see how much smaller *rolling* friction is than *sliding* friction.

Experiment 2.—Arrange a pulley as in Fig. 15, one end of the cord being fixed to a hook, the other end being held in the hand, and attach to the block a mass of from 4 to 6 lbs.

In this experiment we take a larger mass than that in the preceding experiment, because it will at once appear

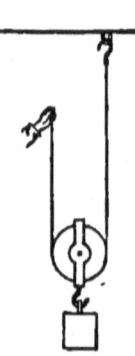

FIG. 15.

that the force required to support the mass in this case, where a *movable* pulley is used, is considerably less than would be required for supporting the mass if held in our hand without the intervention of the cord and pulley; and the difference will be much more noticeable with a large mass than with a small one. Nor shall we fail to observe that less effort is required on our part when our hand applied to the cord moves the mass upwards, than when the same mass is held in our hand and we merely lifted it. Thus it becomes clear that a mechanical advantage is obtained by means of a single movable pulley. But we shall observe at the same time, that while lifting the weight by means of force ap-

CHAP. VI — THE PULLEY 55

plied at the end of the cord, our hand has to move through a larger space upwards than that through which the weight moves upwards. Hence in this case also, as in that of the lever, the mechanical advantage is purchased by loss of speed.

In this experiment our hand lifts not only the mass suspended from the pulley, but also the pulley itself. Hence in our experiments for finding the mechanical advantage gained by a movable pulley or a combination of them, we must not forget to weigh each pulley used and take its weight into consideration, when attempting to find the mechanical advantage of each "system" of pulleys with some exactitude.

Experiment 3.—Arrange a movable and a fixed pulley as in Fig. 16, suspending at P exactly half the sum of the mass suspended at W, and that of the movable pulley. Allow the mass P to descend through a measured distance of exactly 1 foot.

The action of gravity upon the mass P replaces in this experiment the force applied by our hand in the preceding experiment. While P is made to go downwards through 1 foot, the movable pulley with its suspended mass will be seen to rise through exactly one-half of a foot. When left to itself, we shall also find that if P is exactly half the mass of W and the pulley, we shall have equilibrium.

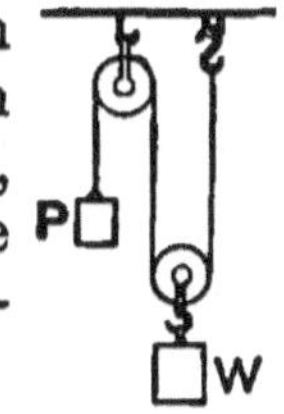

Thus the mechanical advantage of a single movable pulley is 2. But the advantage gained by pulleys admits of being almost indefinitely increased by combination, and a great variety of ingenious devices have been invented with this object.

Fig. 16.

Experiment 4.—Arrange the system of pulleys represented in Fig 17, suspend a mass of 10 or 15 lbs. from

the lower block and apply force at the free end of the cord by means of your hand.

The experiment will at once show that a comparatively small force applied by our hand will be sufficient to balance and to move the suspended mass and pulley. We need not always actually measure the space through which the Power and Resistance move respectively, in order to find the mechanical advantage of any system. A simple consideration will tell us in this case that since the same cord passes here over all six pulleys, fixed and movable, each of the six portions of the cord by which the weight and lower block is suspended must be shortened by the same quantity by which the weight is made to rise; so that the free end of the cord will have to be pulled through a space of 6 inches for every inch through which the weight rises. The mechanical advantage of this system is thus 6. Hence if we replace our hand by a mass equal to $\frac{1}{6}$ of the suspended mass and pulley taken together, we shall find that there is equilibrium.

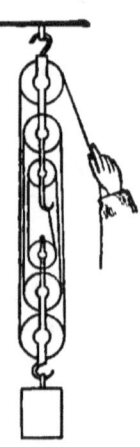

FIG. 17.

Experiment 5.—Arrange the system of pulleys represented in Fig. 18, after carefully noting the weights of the three movable pulleys A, B, C, and suspend a mass of 15 or 20 lbs. from the lower block.

Observe the considerable mechanical advantage gained in this system, first by pulling the free end of the cord with your hand, and then by suspending to it the mass required for equilibrium. This is easily calculated as follows—

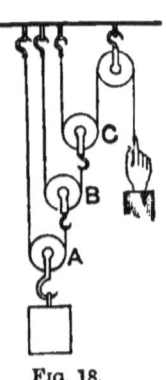

FIG. 18.

When the weight and its pulley rise 1 inch, the cord

on both sides of it will give out 1 inch, and the pulley B will rise 2 inches, while for the same reason C will rise 4 inches, and the hand or the suspended mass will go down 8 inches. Hence we require at the free end $\frac{1}{8}$ of the weight and pulley A, together with $\frac{1}{4}$ of B and $\frac{1}{2}$ of C. Observe carefully that when the system is made to move, B rises twice as much as A, and C twice as much as B, or four times as much as A.

Experiment 6.—**Arrange the system represented in Fig. 19, using weighed pulleys A and B, and a large mass as before.**

This system is of interest, because here the weights of the pulleys assist us in raising the suspended mass, and thus diminish the force which would be required without the pulleys; the reason is that our force is acting on the pulleys in the same direction as gravity; for when the suspended Weight is lifted upwards the work required is done by the Power and the pulleys *jointly,* both doing work in the same sense by moving downwards.

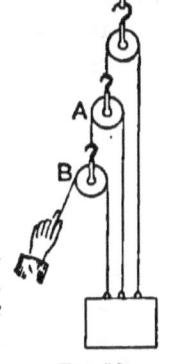

Fig. 19.

If the suspended mass is raised through 1 inch, the cord which passes over the fixed pulley gives out 1 inch, and the movable pulley A descends through 1 inch; hence the cord passing over it gives out 2 inches, besides the 1 inch which it gave out in consequence of the Weight rising 1 inch. The cord over A thus gives out 3 inches, B sinks 3 inches, and the cord over it gives out 6 inches, to which 1 inch must again be added because the Weight rises 1 inch. The Power thus descends through 7 inches altogether, and the mechanical advantage is 7. But if a force be applied equal to the weight of 1 lb. as Power, it will not only support a mass of 7 lbs., but, as has been just shown, the pulley B descends 3 inches, and A 1 inch, when the Weight rises

1 inch, hence the system would support a mass three times that of the pulley B, and one equal to that of the pulley A, if the weights of these pulleys only were made use of to support the suspended mass. It follows that a Power of 1 lb. will support a mass of 7 lbs., together with three times the mass of B and once the mass of A.

Experiment 7.—Arrange two pulleys and a cord as in Fig. 20; suspend a mass of 4 or 5 lbs. and find by actual trial how much must be suspended at P to produce equilibrium.

The mass found in this manner will not be one-half of the suspended mass and pulley as in Experiment 3, but more; nor will P descend through so much as double the space through which the suspended mass ascends, when motion takes place. Comparing our arrangement with that in Fig. 16, we shall not fail to observe that there is an essential difference between the two, although in both we have only one fixed and one movable pulley; but here the portions of the cord which support the pulley and suspended mass are not parallel to one another, they are inclined to the vertical direction, and hence the pull which is transmitted through the cord is not exactly opposite to the direction in which gravity acts. The consequence is that only part of the pull produces an effect in that direction.

We may, however, find by the following geometrical construction, which will be more fully explained and experimentally proved in Part III. of this work, what relation the Power P bears to the supported mass, including the pulley in a case like this. Draw AB vertically as in Fig. 20, and make it as many inches as there are lbs. in the mass W and the pulley; draw BD parallel to the left-

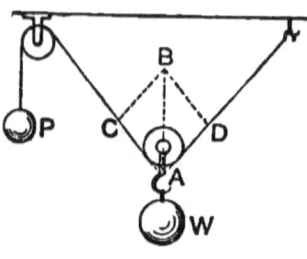

Fig. 20.

hand portion of the cord, and BC parallel to the right-hand portion. It will then be found that the side AC of the figure, which is a parallelogram, contains as many inches as there are lbs. in the mass P, when the system is in equilibrium.

QUESTIONS ON CHAPTER VI

1. Sketch a single fixed pulley and describe its action. Point out in your sketch the *sheave*, the *block*, and the *pivot*.

2. What kind of advantage is gained by employing fixed pulleys? Give examples of their use.

3. What disadvantages would result from using a solid edge or peg instead of the ordinary form of pulley?

4. Which part of the machine is the most essential, the cord or the pulley? Explain fully the reasons for your answer.

5. What similarity can you discover between a lever and a pulley? Where would you look for a point which in the pulley corresponds to the fulcrum of the lever?

6. Sketch a single movable pulley as in Fig. 15, and indicate in your sketch how far the hand has to move upwards, if the space through which the pulley has moved upwards is represented in your drawing by $\frac{1}{2}$ an inch.

7. In the case of Fig. 15 suppose the Power to be 1 lb. and the movable pulley to weigh 1 lb. What would happen if you were to suspend to the block (I) $\frac{1}{2}$ lb.; (II) 1 lb.; (III) 2 lbs?

8. What is the mechanical advantage of a system like that in Fig. 17, if there are 4 movable pulleys in the lower block?

9. Suppose the lower block and pulleys in the system of the last question to weigh 4 lbs., and that we wish to suspend a weight of 36 lbs. to it; what weight is required as Power for producing equilibrium?

10. Suppose the Power and Weight in the system of Fig. 18 to be in equilibrium; what additional weight could be raised if 2 lbs. be added to the Power?

11. Draw the system Fig. 19, but with three movable

pulleys instead of two. What weight could be supported if the Power is 1 lb., and each of the three pulleys weighs 1 lb.?

12. In the system of Fig. 20 the movable pulley weighs 2 lbs., and a weight of 2 lbs. is suspended from it. Find by an *accurate* drawing what Power must be used for equilibrium if the angle between the two portions of the cord is a right angle. Will it be greater or less when the angle is less?

CHAPTER VII

THE INCLINED PLANE—THE SCREW—THE WEDGE

Experiment 1.—Arrange the board used in Chapter II. with its smooth side uppermost and with one end resting on the table, while the other end rests on a firm support half a foot higher. Push the weighted block, used previously, from the bottom of the board to the top with your hand.

You cannot fail to observe at once how small a force is sufficient to carry the block to the top of the board; yet the force is somewhat greater than that required previously when the weighted block, its smooth side resting on the planed side of the level board, was simply pushed from one end of the board to the other. In that case we did

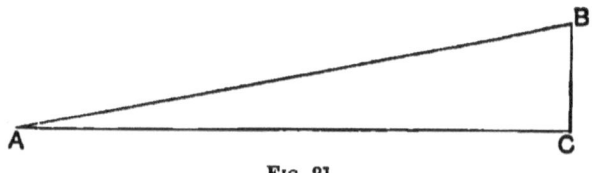

FIG. 21.

work against friction only; in the present experiment we are in addition doing work against gravity, for when the block has reached the raised end of the board we have obviously lifted it vertically upwards through the space by which the raised end of the board is higher than the table.

A flat surface like our board, placed slantingly to the direction in which any mass has to be moved, is called an "Inclined plane." The distance from the bottom to the top, AB in Fig. 21, is called the *length* of the plane, and BC its *height*. The inclined plane is frequently used for raising heavy masses to a certain height by the application of a smaller force than would be required for lifting them to the same height vertically.

Experiment 2. — Lift the weighted block used in Experiment 1 vertically through a space equal to the height of the plane, as measured by a scale.

A much greater force will be required in this experiment than that needed in Experiment 1 for pushing the block upon the plane up to the same height.

Experiment 3.—Increase the height of the plane by making it first about a foot and next a foot and a half; repeat Experiments 1 and 2 in each position of the board.

In the repetitions of Experiment 1 the same mass is moved over the same space, viz. the length of the board, and the effort required will be found to be much greater in that experiment in which the slope of the plane is greater; for in that case the mass is lifted upwards against gravity to a greater vertical height. In the repetitions of Experiment 2 the mass is raised vertically to different heights without using the inclined plane. Now the work done, when a mass is lifted vertically upwards is, as we have already learnt, equal to the product of the resistance into the space through which it is overcome; in this case to the weight of the mass into the height of the plane. Again, if we consider Experiment 1 or any of its repetitions we see that, leaving friction out of account for the moment, the only effect produced in each experiment is that the mass is raised to a certain height above the table. The work done in each experiment is thus clearly the same as in each of the repetitions of Experiment 2, when the mass is lifted straight up to heights corresponding to those of the first series of experiments. But in the first series the mass is raised by force being applied along the whole length of the plane; in the second series force is applied only through a space equal to the height of the plane. It follows that the force required in the first case must be by so much less than that in the second case, as the height of the plane is less in proportion than the length of the plane. Hence, as in the similar cases of the lever and pulley, a mechanical advantage is gained in either supporting or moving a mass on an inclined plane, which is equal to the length of the plane divided by its height.

We know from previous experiments that the block will not slide but will rest on our board if one end be raised, provided that the inclination does not exceed a certain limit. Within that particular angle of inclination friction alone is thus sufficient to maintain a mass at rest upon an inclined plane and no other force is required. As a consequence of this we cannot find by actual experiment the relation between the Power and Resistance on the inclined plane so easily and accurately as in the case of the lever and pulley, because friction assists the Power in this machine to a certain extent, which is of course different for each different apparatus used, and is in any case not easily determined by simple experiments.

Experiment 4.—Make the height BC (Fig. 22) as great as convenient, fix a pulley near the top B, and attach a string to the mass R, so as to be parallel to the plane and to pass over the pulley. Adjust the mass P until the mass R is kept at rest on the plane.

By making the plane as much inclined to the horizontal direction as possible, we diminish friction and therefore its influence upon our experiment as much as possible; for the pressure which our mass exerts on the plane is the greater the more horizontal the plane is, and the less the more vertical the plane, and we have learnt that friction diminishes when the pressure between the rubbing surfaces decreases.

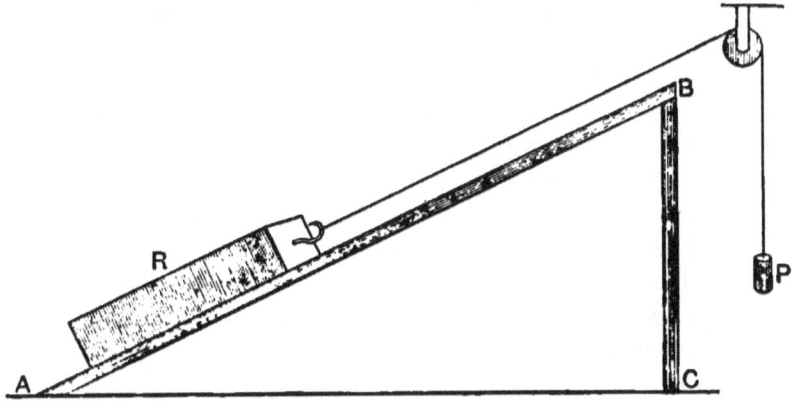

Fig. 22.

We shall thus find that the weight of P will be to that of R very nearly in the same proportion as the height of the plane is to its length. A little consideration will also tell us that if the block R is made to rise from the bottom of the plane to its top, that is through AB, by making P a little larger than is required for balancing R, R has really been lifted against gravity through the vertical height BC, while gravity has done the same amount of work upon P, the smaller mass being moved through a proportionally larger vertical space which is equal to the length of the plane, AB.

The commonest example of an inclined plane is a road which is not level, and the relation between the Power and the Resistance which has been established by the preceding experiments holds good with reference to the loads usually drawn on wheels upon roads, and the forces exerted by horses, engines, etc., employed to impel them. In practice the inclination of the road is estimated by the height corresponding to some definite length; thus when a road is said to rise, or to have a "gradient" of, one foot in fifteen, or one yard in twenty, the meaning is in the first case that if a length of fifteen feet on the road be measured off, it represents an inclined plane of which the height is one foot and the length fifteen feet; in the second case similarly, the vertical height to which a load is lifted is one yard, when the power has passed over a slanting length of twenty yards. This convenient mode of estimating the inclination of roads clearly gives at once the relation between Power and Resistance, and therefore the mechanical advantage. For example, on a gradient of "1 in 60" a Power of 1 will be sufficient to sustain a Resistance of 60. On railways the gradient seldom exceeds 1 in 60; on ordinary roads where the friction is so much greater the gradient is sometimes as much as 1 in 7.

Other familiar practical applications of the inclined plane are: a flight of steps, and the two long poles used for rolling a barrel up into a waggon.

Experiment 5.—Place the wooden block used in Experiment 1, weighted by a very heavy mass, upon the table. First lift the block and the mass together with your hands through a few inches; then apply your hand at the thicker end of the wedge-shaped piece of wood ABC (Fig. 23), so as to push the thin edge along between the block and the table as far as possible.

In this experiment it is not the mass which is moved up the plane, but the inclined plane itself is urged under

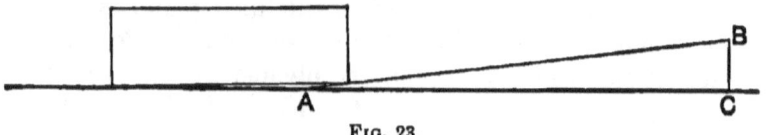

Fig. 23.

the mass by force applied to the back of the plane, BC.

The force in this case does not act in a direction parallel to the length of the plane but parallel to its "base" AB. The work done in this way will be that the block is lifted against gravity through a vertical height equal to BC, if the whole of the plane has been propelled beneath the block. We have therefore the following relation—

$$\left. \begin{array}{c} \text{Force applied} \\ \text{multiplied by base of plane} \end{array} \right\} = \left\{ \begin{array}{c} \text{Resistance} \\ \text{multiplied by height of plane,} \end{array} \right.$$

hence,

$$\text{Mechanical advantage} = \frac{\text{Base of plane}}{\text{Height of plane}}.$$

We shall thus see why our effort in pushing the plane under the mass is found to be considerably less than that required for lifting the block vertically.

When the inclined plane is applied in this manner it is called a *Wedge*. In most cases the wedge is formed of two inclined planes placed base to base as in Fig. 24, which represents an ordinary wedge as used for splitting wood. The thin edge is inserted in a cut previously made in the wood, and force being applied at the back or broad end of the wedge by pressure or a blow, the wedge is driven onward through a certain space and the portions of wood on both sides are farther separated. Thus, in this case, the work done by ourselves is not done against gravity, but against cohesion.

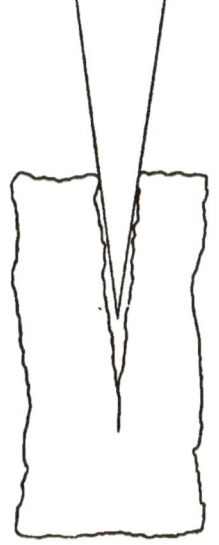

Fig. 24.

Wedges are frequently used where great resistance is to be overcome through a comparatively small space. Thus ships are raised in docks by wedges driven under their keels; and instances have occurred in which the wedge has been used to restore a tottering building to its perpendicular position. All cutting and piercing instruments, such as knives, razors, chisels, nails, pins, needles, swords, bayonets, etc., are wedges.

Experiment 6.—Cut a right-angled triangle, ABC (Fig. 25), out of writing paper, draw a black line near the edge AB, and wrap it firmly round a cylindrical body, such as a stout piece of pencil, 2 or 3 inches long.

Our piece of paper represents in fact an inclined plane, of which BC is the height, AB the length, and AC the base. When the paper is rolled round the cylindrical

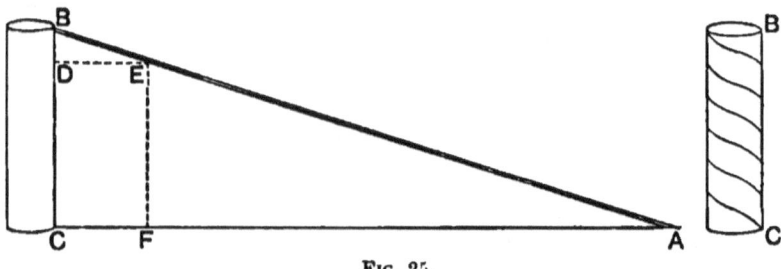

FIG. 25.

body, the black line appears as a spiral line around it, while the pencil has the well-known appearance of a screw, of which the black line represents what is called the "thread."

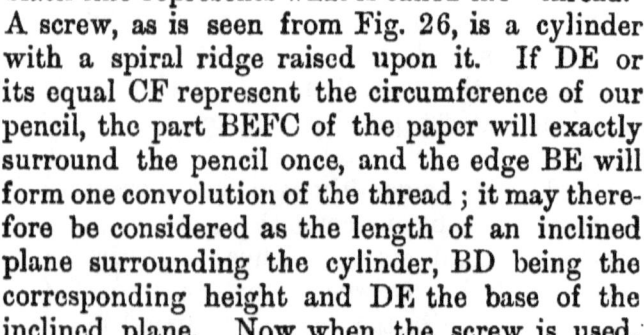

FIG. 26.

A screw, as is seen from Fig. 26, is a cylinder with a spiral ridge raised upon it. If DE or its equal CF represent the circumference of our pencil, the part BEFC of the paper will exactly surround the pencil once, and the edge BE will form one convolution of the thread; it may therefore be considered as the length of an inclined plane surrounding the cylinder, BD being the corresponding height and DE the base of the inclined plane. Now when the screw is used, the force applied acts parallel to the base DE of this plane, because it acts at right angles to the cylinder BC, as we can see at once, when turning any screw. It

CHAP. VII — INCLINED PLANE—SCREW—WEDGE

follows from this and the principles already proved that the Power will be to the Resistance as the length BD is to DE; but BD is evidently the distance between the successive positions of the thread as it winds round the cylinder. Hence in general we may state that in the screw—

$$\text{The mechanical advantage} = \frac{\text{Distance through which the Power moves}}{\text{Distance through which the Resistance moves}}.$$

and taking only the work done during one revolution into consideration, while the force is supposed to be applied at the screw-cylinder itself—

$$\text{The mechanical advantage} = \frac{\text{Circumference of screw}}{\text{Distance between two successive threads}}.$$

It follows from this relation that the mechanical advantage becomes greater when the distance between the threads, which is commonly called the *pitch* of the screw, is lessened, or in other words, the finer the thread is the more powerful the machine will be. Further, the distance through which the power moves may be considerably increased by applying the force not immediately to the screw itself, but to a solid bar firmly attached to the screw, as shown in Fig. 27. The mechanical advantage in this screw is thus made greater as many times as the length AB of the attached bar is greater than the radius ab of the screw-cylinder itself. We may then express the whole mechanical advantage in this case thus—

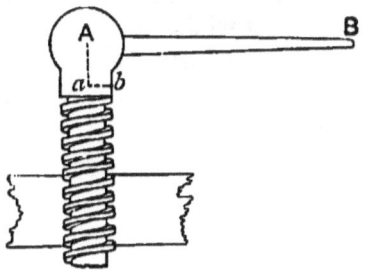

Fig. 27.

$$\text{Mechanical advantage} = \frac{\text{Circumference of circle through which the point B moves}}{\text{Distance between successive threads}}.$$

To use the screw it is necessary to have a hollow

cylinder with a groove cut on the inside of it, as seen in Fig. 27, so that the thread of the screw exactly fits into it. This hollow cylinder is called the "nut." If the nut is fixed while the screw is turned, the screw passes through the nut; the converse takes place when the screw is fixed and the nut is movable.

Screws like those in Figs. 26 and 27, in which each turn of the thread appears higher on the right-hand side than on the left when the screw is held upright, are called right-handed screws; when the left is higher, as the edge of the paper in Fig. 25, it forms a left-handed screw.

There are many applications of the screw in the arts; it is used, for example, where liquids or juices are to be squeezed out from solid bodies; in coining, where the impression of a die is to be made upon a piece of metal; in printing and book-binding; in compressing soft and light materials, such as cotton, so as to reduce them to a convenient bulk for transportation; and in a great variety of instances for securely fastening, as in ordinary screw nails.

QUESTIONS ON CHAPTER VII

1. A heavy mass is kept at rest on an inclined plane by a force acting parallel to the length of the plane. Indicate by straight lines the direction (I) of that force; (II) of the friction which acts between the mass and the plane; (III) of gravity; (IV) of the pressure on the plane.

2. Which of the three forces I, II, IV, in the preceding question is most likely to be the greatest, and which the smallest, if the inclination of the plane to the horizon is (*a*) very small; (*b*) very large? Why is force III left out of the comparison?

3. What is meant by a gradient of 1 in 40? What Power is required to keep a carriage weighing 2 tons at rest upon a road having that gradient, if there is no friction?

4. Give as many examples as you can of practical uses of (I) the inclined plane; (II) the wedge; (III) the screw.

5. An inclined plane is 6 feet long and 1 foot high; if friction can be neglected, find how much work is done by a power of 10 lbs. acting along the plane and pushing a mass from the bottom of the plane to the top. What is the magnitude of the mass thus raised? Explain your calculations.

6. Explain in what respects the wedge and the screw are similar to the inclined plane.

7. A wedge is 1 foot long and its greatest thickness is 2 inches. What is its mechanical advantage? Explain your calculation.

8. The pitch of a screw is 1 inch; the length of the handle by which it is turned is 15 inches. If the power applied to the end of the handle is 12 lbs. what resistance can be overcome by turning the screw if friction is neglected?

CHAPTER VIII

MOMENTUM

Experiment 1.—Place a small square block of wood close to the left-hand edge of a table, and strike it a moderately strong blow with a hammer, thus causing it to fly off the table.

The hammer is set in motion by muscular force applied to it; it moves through a certain distance and strikes the block, which takes up, as it were, the motion of the hammer, and begins to move in the same direction as that in which the hammer itself would have continued to move if its own motion had not been transferred to the block. Our experiment is so arranged that the friction between the block and the table scarcely interferes with the free motion of the block, which however is taking place under the action of *two* forces; for gravity and the hammer are acting upon it at the same time. Now if the blow of the hammer were applied in a horizontal direction, and no other force were acting upon the block, there would be no reason why the block should move in any other but the hori-

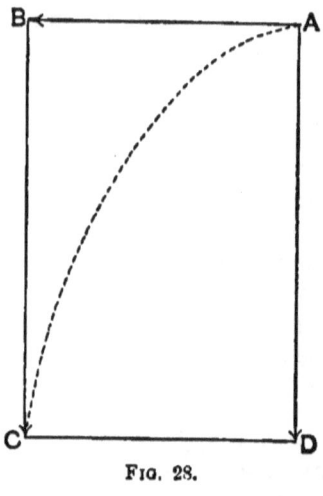

Fig. 28.

zontal direction. Again, as the block leaves the support of the table it is free to move under the action of gravity, which, as we know, causes bodies to move vertically downwards. Our experiment shows that the block thus urged on by two forces, one represented by the horizontal line AB, Fig. 28, the other by the vertical line AD, moves in a direction intermediate between the two, as indicated by the dotted line AC, which, if we watch the motion of the block very carefully, will be seen to be not a straight line, but a curve, the form of which will depend upon the relative magnitude of the two forces. Thus if the force which produces horizontal motion is made greater, while gravity of course remains the same, the curve AC in Fig. 29 will represent the actual motion more nearly than

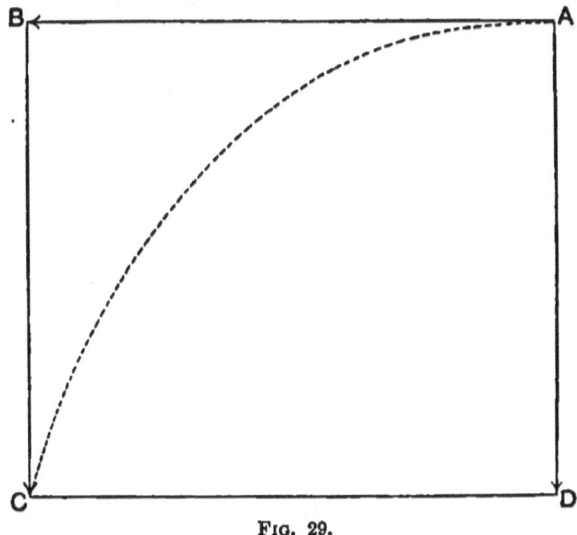

Fig. 29.

AC in Fig. 28. For the present, however, our attention will be chiefly directed to other results founded upon this experiment and those which succeed it.

The effect of the blow which was given was to cause the block to move with a certain *speed*. The notion which is in our mind when we speak of a moving body having a certain speed, is that the body is capable of passing over a certain space in a certain time. The speed of a moving body, or as it is called, its *velocity*, is generally measured by the space which a body passes over in a second of time. Thus if suitable instruments for measuring space and time were at our disposal in this experiment, and we found that the block has moved over 6 feet in two seconds, we might say that the velocity with which we caused the block to move was 3 feet per second. We may neglect for our present purpose the very important fact that the velocity of the block is in our experiment by no means the same throughout its motion, it does not pass over equal spaces during successive equal intervals of time. In other words, its velocity is not *uniform*; but this will not interfere with the correctness of our general conclusions.

Experiment 2.—Repeat the preceding experiment three or four times, using the same block, but gradually increasing the force applied to the hammer in the successive experiments.

We shall find, as we should naturally expect, that the block is driven farther from the edge of the table in each successive experiment. But the essential point, brought out even by a rough observation, is that the speed with which the block moves is becoming greater and greater as the force which is the cause of its flight is increased more and more. In Part III. of this Course experiments will be described which will enable us to measure very exactly the forces producing motion, the masses which are caused to move by these forces, and the spaces passed over in each successive second, and we shall then be able to establish by accurate experiments that *if forces of different magnitudes are applied to a mass so as to produce motion, the velocity with which the mass moves increases or diminishes in the same proportion as the force increases or diminishes.* Thus if a mass of 1 lb. is set in motion by some force so as to move with a velocity of 1 foot per second, a force of twice the magnitude will cause the same mass to move with a velocity of 2 feet per second; a force three times as great will produce a velocity of 3 feet per second in the same mass, and so on.

Experiment 3.—Repeat the experiments again, but use three blocks of different mass in succession, applying as nearly as possible the same force in each experiment.

We shall find that the block of smallest mass will move over the greatest distance before it reaches the ground, that which is next in mass will pass over a less distance, and that of largest mass over the least distance. By observing carefully the curve described by each mass, or, as it is called, its *path*, we shall find that the path of the smallest mass will resemble that in Fig. 29, while that of the largest mass approaches more to the form of the curve in Fig. 28. This shows that the same force produces a greater effect in its own direction when acting upon a smaller mass than when acting upon a larger mass. Finally, it will be observed without any room for doubt that the smaller the mass the more speedily it passes over a given space during its flight, this becomes especially apparent during the first portion of the time occupied by the motion of the different blocks used in our experiments.

In this, as in the preceding case, more accurate experiments prove still more exactly that *the greater the mass which the same force sets in motion the less is the velocity produced*; or, as this law is usually stated, when equal forces act upon different masses the velocities produced are inversely as the masses. Thus the force which acts upon a mass of 1 lb. and produces a velocity of 6 feet per second, will, when acting upon a mass of 2 lbs., produce a velocity of

3 feet per second, upon 3 lbs. a velocity of 2 feet per second, and so on.

The preceding experiments lead to very important consequences. In the first place, if means are at our disposal for measuring accurately the rate at which a moving body passes though space, that is, its velocity, we are now enabled to compare the magnitude of various forces. For we might agree upon some unit of mass, say 1 lb., and measure accurately the velocity produced in each case if these various forces are allowed to set this standard mass in motion: the greater the velocity the greater will be the force.

Next, we are provided with an accurate means of comparing different masses with each other; for if this is our object we may agree upon some unit of force, that force, for example, which, acting upon the unit of mass as previously defined, produces unit of velocity, say 1 foot per second, and hence express the various masses, which we wish to compare, in terms of the unit mass, by observing the velocities produced when each of the masses to be compared is set in motion by our unit of force.

This mode of proceeding would, however, present a great many difficulties in practice, though its theoretical consideration as stated here will no doubt help us to a clearer insight into the connection between force, mass, and velocity than we had without it, and before making our experiments. One great difficulty must, however, be stated at once. In all these experiments the force acted instantaneously, and the original cause of motion, having produced its effect, ceased to act further upon the mass after it had once begun to move. Force was applied to the hammer, which caused the hammer to move; its motion was then in great part given up to the wooden block by a blow which lasted only an infinitely small interval of time. But, as a rule, forces do not act thus, and their duration is not confined to such short intervals.

A piece of iron near a magnet, a piece of paper or of rubbed glass near rubbed sealing-wax, are acted upon by magnetic and electric forces during a measurable interval of time; a body falling under the action of gravity is acted on by gravity, not only before and during its fall, but even after it; in fact, unceasingly. We must admit that these cases differ in many respects from the mode in which force acted in our experiments; nevertheless we shall see further on that the mode of dependence of the velocity acquired upon the force causing motion and the mass moved which has been stated above, is also found to hold in the case of forces which continue to act upon the moving body during its motion.

Experiment 4.—Take a small cannon resting on wheels, and load it with powder and a bullet of suitable size. Fire it, so that the shot may hit a solid block of wood placed a few inches from the mouth of the barrel.

When the charge of powder is fired, chemical action takes place, and instead of the solid powder we obtain a mixture of different gases of much larger volume, and exerting a pressure which is very considerable in all directions. It is this force which produces the two effects which are observable—first, that the bullet strikes the block, enters it, and probably (if the friction between the block and the table does not prevent it) moves the block through a short distance or overturns it; secondly, that the little cannon itself will be seen to move backwards on its wheels through a certain space, a movement which is technically called the "recoil" of the gun. Now the force exerted upon the mass of the bullet and that upon the cannon may be considered to be equal, hence the effects produced must be equal also. If the mass of the cannon were equal to that of the bullet, we should from the preceding experiments expect the cannon to move backwards with the same velocity with which the bullet is projected in the opposite direction; but the cannon has a much larger mass, hence its velocity will be as many times less than that of the bullet as its mass is greater. Thus the force which acts in this case could be numerically expressed either by the product of the mass of the bullet and its velocity, or by the product of the mass of the cannon and its velocity, both these products being equal, because the force which produced each motion was the same. To this product of the mass and the velocity of a moving body the name *momentum* has been given. If a body of unit mass, that is 1 lb., moves with a velocity of 1 foot per second, it is said to have unit momentum, and we may take the force which would produce unit momentum as our unit force. If a body has a mass of 200 units, and moves with a velocity of 1 foot per second, the momentum is 200, and clearly the force producing motion must have been 200 times greater than the unit force; the same will be the case if a mass of 1 lb. moves with a velocity of 200 feet per second, or a mass of 10 lbs. with a velocity of 20 feet per second, or a mass of 50 lbs. with a velocity of 4 feet per second. In all these cases the momenta of the moving bodies are the same, and provided that in each of these cases the forces acted during an infinitely small time, as did the blow given by the hammer or the explosion of the gunpowder, the momentum produced in each case is an accurate measure of the force which produced the motion. But if a force constantly acts through a definite interval of time, its effect upon a given body will depend upon the length of time during which it has acted, and it will appear at once that a weak force acting for a long time may be capable of producing effects equal to those of a greater force acting during a short time; hence for a more complete consideration of the effects of forces upon masses the element of time will have to be taken into account, and this will be considered later in its proper place.

Experiment 5.—Dismount the cannon from its carriage; suspend the former so that it is horizontal, and opposite to its mouth, and about 5 or 6 inches from it, suspend a small lead target, of about the same weight as the cannon. Load and fire again.

The object of this mode of suspension is to diminish friction as much as possible, and to place the cannon and the target in similar and therefore comparable circumstances. They are both now like pendulums, in which the bobs have equal weights, and their mode of suspension is very much alike. When the cannon is fired the two pendulums will be set in motion, and, as will be seen at once, in opposite directions, the gun moving to the side opposite to that in which the bullet is fired, while the target which receives the shot and its motion moves in the direction in which the shot moves. Thus the two bodies will alternately recede from one another, approach each other, recede again, and so on; and as their motion is as nearly as possible the same, we may conclude that the forces which caused the motion were equal. Having thus convinced ourselves that the impressed forces in Experiment 4 were equal, we cannot doubt that their effects were equal, and that these effects can be correctly represented by the *quantity of motion* produced, which is nothing else but the product of the mass moved and its velocity, that is, the momentum of the moving body.

QUESTIONS ON CHAPTER VIII

1. Explain, as fully as possible, the meaning of the term *velocity*.

2. When a body passing over one foot in one second is said to have unit velocity, what is meant when we say of a body that its velocity is 48?

3. A railway train passes over 40 miles in an hour; what is its velocity in feet per second?

4. The velocity of a racehorse is 36 feet per second, what time will it require to run over a course of 3 miles?

5. Explain fully what is meant by the term *momentum*.

6. A cannon ball weighing 2 lbs. is moving at the rate of 1200 feet per second; what is its momentum?

7. A cricket ball weighing $\frac{1}{3}$ lb. is bowled with a velocity of 12 feet per second, and hit with a velocity of 21 feet per second. What is its momentum before it is hit and after?

8. A leaden ball weighing 3 lbs. moves with a velocity of 10 feet per second, and strikes full against another ball which weighs 2 lbs., and moves towards it with a velocity of 15 feet per second, what will happen when they meet? What would have happened if they had been moving in the same direction and the second ball had overtaken the first?

9. Two different forces act for one second upon the same mass, one produces a velocity of 12, the other of 24. Compare the magnitudes of the two forces.

10. The same force acts for the same time upon two different masses, producing in the first a velocity of 12, in the second of 24. Which mass is greater? How many times greater is the one than the other?

11. A greyhound weighing 30 lbs. and running at the rate of 35 feet per second, is chasing a hare weighing 7 lbs. and running at the rate of 30 feet per second. Compare the momenta of the two, and explain why the hare, although running slower, has a good chance of escape.

12. A shell is flying along with a velocity of 600 feet per second, and explodes, forming two parts of equal weight, one of which is driven forward in the same direction with an additional velocity of 600 feet per second. What will happen to the other half? Calculate the momenta of the shell before the explosion and of its parts after it, if the whole shell had a mass of 20 lbs.

CHAPTER IX

ENERGY

Experiment 1.—Repeat Experiment 4 of the preceding chapter. Immediately after the bullet has entered the wooden block, insert the bulb of a small thermometer into the hole made by it.

We know that the bullet would have been projected to a considerable distance from the gun if the wooden block had not been in the way. The motion of the bullet being stopped, it is, as we have already learnt, converted in great part into heat, and hence we shall find that the thermometer, if it is sufficiently delicate, will sensibly rise. But besides heating the target to a certain extent, the bullet enters the wood to a certain depth, and thus overcomes the cohesion of the solid particles. Thus the bullet is proved to possess energy; it is capable of doing work, and the work done, as we see, is mainly spent in splitting some of the wood, and in heating it. If we were to repeat Experiment 5 in the preceding chapter, using an iron target, we should find the heating effect greater, and quite easily to be detected by placing the tip of the finger quickly upon the spot where the bullet hit the target; and if we touched the bullet itself as soon as it dropped after impact with the target, we should find it quite warm. The iron presents greater resistance than either wood or lead, hence less mechanical work is done by the bullet when hitting the iron target than in the other cases, consequently more energy of motion is converted into heat.

Next let us turn our attention to the gun itself, which has recoiled to the distance of several inches, if the table on which the experiment is performed is fairly smooth. If the table were rough, or the gun did not rest on wheels, the recoil would amount to very little. Still some work has been accomplished by the gun also, but compared with that done by the bullet it is quite insignificant. The heating effect, though it exists to a small extent, cannot be detected by placing either our hand or a thermometer upon the track on the table along which the gun has moved, or upon the rim of the wheels. Indeed the only resistance which presented itself to the gun when momentum was given to it by the pressure of the gases was the friction between it and the table; and this friction is so small that we are inclined to expect the gun to roll backwards through several feet at least. But nothing of the kind happens. If we had placed a wooden block in the path of the recoiling gun before firing, the gun would not, like the bullet, have been driven into the wood. Thus the general conclusion to which our experiment leads is, that although two moving masses may have the same momentum, as our bullet and our gun, which have been proved to possess as nearly as possible the same momenta, they may yet differ as widely as possible in their capacity for doing work, that is, in the amount of energy which each of them possesses.

Both masses have the same momentum; they have both been set in motion by equal forces, but one mass being a certain number of times greater than the other, its velocity is the same number of times less than the velocity of the smaller mass, and seeing that more work is actually done by the smaller mass, we are compelled to seek for an explanation of the observed difference in energy by looking to the greater velocity of the smaller of the two moving masses as the source of its greater energy.

Experiment 2.—Load the cannon with a well-fitting bullet and a very small weighed-out quantity of powder. Use some soft wood, or still better, a number of thin laths of pinewood placed close upon one another, as a target. Discharge the cannon, its mouth being about 3 inches from the target.

When the cannon is discharged the shot leaves the mouth of the gun with a definite velocity, in virtue of which it is able to do work. We are justified in assuming that the depth to which the shot enters the wood is roughly a measure of the work done, that is, of its energy. We may further, for the purpose of the present simple comparison, take the mass of our shot as unit mass, its velocity as unit velocity, and its energy as unit energy. We ascertain how much work has been done upon the wood by probing carefully and measuring how deeply the shot has entered the wood, or how many of the superposed thin laths have been penetrated by it, and make a note of the result. We must bear in mind the fact that the bullet came to complete rest as soon as it had entered the wood for some distance; that is, disregarding for the present the energy expended in heating the wood, or in any other way, of which we shall take cognizance further on, we must remember that the bullet has expended its whole energy while penetrating the wood to a certain depth which has been measured by us.

Experiment 3.—Load the cannon with the same size of bullet several times in succession, but increase the charge of gunpowder, so as to make it first $1\frac{1}{2}$, 2, and $2\frac{1}{2}$ times the charge used in the last experiment.

We shall be able to ascertain that the bullet penetrates each time deeper into the wood, and that, consequently, more energy is possessed by the bullet and expended in penetrating into the wood when the charge is increased. We are justified in assuming that this energy, since the moving masses are in each experiment as nearly the same as we can make them, must depend on the velocity with which the bullet is moving. We may certainly assume that a greater charge of gunpowder when exploding will effect a greater pressure upon a solid mass upon which this pressure is acting than a smaller charge, but beyond the general conclusion that the velocity with which our bullet leaves the mouth of the cannon in the first experiment is less than in the succeeding experiments, no definite numerical law can so far be established from these experiments. The energy of the bullet increases, as is seen by the greater work done by it; it has the same mass, hence the energy must increase when the velocity increases. But in what manner does the energy depend upon the velocity of the moving body?

Experiment 4. — Fill an ordinary water pail with moist clay, and drop a heavy leaden bullet upon the clay first from a height of 2 feet, next from a height of 8 feet. Measure each time to what depth the bullet has penetrated into the clay.

We shall find that the bullet dropped from a height of 8 feet penetrates to four times the depth that it did when falling through 2 feet.

The difficulty which presented itself to us in the previous experiment is chiefly this, that we have no means of ascertaining the exact velocity with which in any of the experiments the bullet either leaves the mouth of the cannon or arrives at the target. But in the present experiment we may be guided in our conclusions by observed facts, which we shall ourselves establish in Part III. It will be shown there that if a body falls under the action of gravity, the velocity which it has acquired at the end of the first second of its fall is 32 feet per second; at the end of two seconds it is 64 feet per second; at the end of three seconds 69 feet per second, and so on, and from what has already been proved about the action between the same force on different masses, and different forces on the same mass, we know that every falling body, whatever its mass, will acquire the same velocities in 1, 2, 3, etc. seconds.

Next, if a body falls for one second, its velocity, which was equal to zero at the moment when it was dropped, increases uniformly during the second until it is 32 feet per second, during which time the force of gravity has been acting upon it without break or sensible alteration. We may thus conclude that the space which the falling body described in the first second of its fall is the same as that which a body would describe moving uniformly with a velocity of 16 feet per second, which is the *average* velocity of the falling body. Hence we infer that the space through which a freely falling body passes in the first second is 16 feet. After two seconds the body has a velocity of 64 feet per second; the average velocity throughout its motion is 32 feet per second; but it moved two seconds, hence it describes 64 feet in the first two seconds. That is while the velocity became doubled, the spaces passed over became quadrupled, or suppose a falling body which passes over the space of 2 feet, as in our experiment, to have acquired unit velocity, then in falling over four times that space, 8 feet, it will have acquired twice the unit velocity.

Our bullet thus has double the velocity when reaching the clay in the second experiment compared with that which it had when arriving at the clay in the first experiment, because it had passed through a space four times as great; but it did four times as much work as it did in the first experiment, hence follows the important law that *the energy of a mass in motion is proportional to the square of the velocity.*

We may suppose the bullet used in these two experiments to represent unit of mass, moving in the first experiment with unit of velocity when reaching the clay. The energy may then also be represented by unity. In the second experiment, the mass being the same, but the velocity double, the energy is four units. Hence we represent the energy numerically by 4. On the other hand, the momentum of the bullet would be in the first experiment = 1, in the second = 2. The fact that the energy increases as the square of the velocity explains why, of two masses having the same momentum, the one with the greater velocity can accomplish so much more work. Thus, suppose we assign to our little cannon used in previous experiments a mass of 100 units, while the bullet represents 1 unit, and let the bullet be projected with a velocity of 100 units, then to compare the momenta and energies of cannon and bullets we have—

	Bullet.	Cannon.
Mass	1	100
Velocity	100	1
Momentum	$1 \times 100 = 100$	$100 \times 1 = 100$
Energy	$10{,}000 \; (= 1 \times 100 \times 100)$	$100 \; (= 100 \times 1 \times 1)$

Thus the bullet can do 100 times as much work as the cannon, though originally the same force was used to set in motion either of the two bodies, or rather two systems of bodies, namely the cannon on the one hand, and on the other the bullet, target, and the air all round; of which more further on.

Experiment 5.—Drop from a height of 1 foot first a one-pound weight, and then one of 2 lbs. upon tacks stuck side by side into a block of wood, as in Experiment 1, Chapter III.

Both these masses are set in motion by gravity as soon as we withdraw the support of our hand; but the force which acts on the mass of two pounds is twice as great as that acting on the mass of 1 lb. On the other hand, the mass set in motion is just twice as great in the case of the 2 lbs. as in that of the 1 lb., and hence the velocities with which the two bodies move will, as they are dropped from equal heights, be throughout their fall the same at any two points equally distant from their point of starting. Now the result of the experiment proves that the mass of 2 lbs. drives the tack twice as deep into the wood as the mass of 1 lb. The two bodies, however, have the same velocities, while the mass of one is twice that of the other; hence we conclude that *the energy of a moving body is proportional to the mass of the body.*

The results of the experiments made in this and the preceding chapter, and also in Chapter III. have prepared us for entering upon a closer acquaintance with the various forms of energy which exist in nature, and to the consideration of which the remainder of this part of the present work will be devoted. It will, however, be desirable to glance once more at the general laws which have thus far been obtained.

We have seen that all motions of bodies are ascribed to causes which are called forces, and we have already convinced ourselves in Part I. that bodies may be set in motion by a variety of such causes, as for example by gravity, chemical affinity, elasticity, capillarity, heat, electricity, and others. If any of these forces is supposed to act only for an infinitely small interval of time, as was in our experiments the case with a blow, or the sudden pressure of a gas, then the effect is correctly measured by the momentum of the body set in motion, that is, by the product of the mass and velocity. But even in such a case time is actually required for the action of the force; the body which imparts the blow cannot stop its own motion at the very instant when it touches the other body which is set in motion, it moves for a very short time with it, and however small this interval of time may be, it must be admitted that the force applied, or part of it, continues to act during the small interval. Similarly in the case of the pressure of the gases generated by the explosion of powder. While the ball or shot is passing from one end of the barrel to the other, it is continually acted on by the pressure of the gases, and however small a fraction of a second this may be, still we cannot properly speak of an instantaneous action. Time, and its influence upon the effect of a force must therefore form a part of a complete investigation of these effects; and we have only postponed our experiments on this part of our study because we are at present chiefly aiming at making a better acquaintance with the various forms of energy in nature, and we can for a time dispense with the results of those experiments.

We live in a world of work, and for all the purposes of life the most practically important part of the investigation of the properties and effects of the various forces in nature is to determine how much work can be done under given circumstances by a body in motion and subjected to them. Considerations on the subject of the forces themselves, though they are the original causes of all motions, or on that of the momenta which they can produce, are subsidiary to this. The actual tangible result in the shape of work which can be got out of a body having energy in it, is invariably looked for in all practical applications, and it is for this reason that such forces as gravity, chemical affinity, heat, and electricity are of such far-reaching importance to us; other forces, for example capillarity and elasticity, stand far below them in the amount of work which can be obtained by their agency; while the force of friction concerns us chiefly when we consider the amount of useful work which is turned away from its legitimate goal by the action of this force.

Now, in considering what amount of work can be obtained from a moving mass, the first thing which our previous experiments suggest is, that work can only be done while energy is being transferred from one body to another. A cannon ball does no work while it is moving through the air, except the comparatively very slight amount done in pushing aside the particles of air in its way; it is only when it comes into contact with something which will take some of its energy out of it, when it encounters some *resistance* in fact, that we observe that work is done. The portion of the target struck by the ball takes up some of its energy in the shape of actual motion, particles of wood, iron, or stone are set in motion and become themselves capable of doing work, and if the ball is finally stopped, as well as the splinters, etc. which have been projected, the energy has assumed a different form, that of heat, which in this new form may, by proper contrivances, be rendered still capable of doing work if the heated matter can be made to act suitably upon material particles which are at a lower temperature than itself. Similarly a billiard-ball, set in motion, loses little energy in travelling over the table, but when it strikes against another ball its energy is transferred to that ball, which immediately moves on at almost the same rate as the first ball, while the latter stops, owing to the loss of its energy, and can do no more work until struck again by the cue.

In the next place the preceding experiments present us with a very interesting example of a body possessing energy without being in motion, the gunpowder before the explosion bearing a similar relation to the energy displayed by the gases to which it gives rise after the explosion as the raised weight in the experiments of Chapter III. bore to the energy which could be obtained from it before it was dropped and allowed to do work in virtue of its motion. Work is required to lift a weight, and in consequence of being raised it possesses energy of position, which enables it to perform exactly the same amount of work which had been originally expended to raise it; similarly work was required to separate the sulphur and charcoal contained in the powder from the oxygen with which these elements are usually combined, and thus to place them in a favourable position to combine with the oxygen presented to them in the nitre of the powder, a body rich in oxygen, as soon as a spark gives the necessary start to the motion. This kind of energy, possessed by a body not in motion but due to position, frequently also called *potential* energy (from *potentia*, ability), points strikingly to the difference between momentum and energy, for a body can only have momentum when it is moving, but it may have energy when motionless.

Finally, this question now suggests itself, how much work in foot-pounds can actually be done by a body in motion? We have seen that the energy is proportional to the mass, and to the square of the velocity with which the mass moves. Thus if a mass of 1 lb. moves with the velocity of 1 foot per second, and another mass of 2 lbs. moves with a velocity of 2 feet per second, we are perfectly justified in saying that the second body has twice the energy of the first, because its mass is double, and further four times the energy of the first because its velocity is double, hence altogether its energy is eight times as great as that of the first body; but it is incorrect to suppose that the work which the first body can do is 1 foot-pound, or that the work which the second body can do is 8 foot-pounds.

It is plain that this is not the case, for a mass of 1 lb. acted upon by its own weight until it has acquired a velocity of 1 foot per second has not to fall through 1 foot for the purpose, and therefore does not contain 1 foot-pound of work.

It will be shown in Part III. that the actual distance through which a body has fallen from a state of rest when a velocity of 1 foot per second is acquired is $\frac{1}{64}$ foot; it follows, that a body whose mass is 1 lb., and which is moving upwards with a velocity of 1 foot per second, will rise $\frac{1}{64}$ foot in consequence of having this velocity, and hence it will contain $\frac{1}{64}$ foot-pound of energy. If, then, a moving body has a mass of M lbs. its energy will be M times as great, and if it has a velocity of v feet per second its energy will, as we have seen, be v^2 times as great in addition.

Hence we have, if M is the mass of a moving body in lbs., and v its velocity in feet per second, the very important formula—

$$\text{Energy} = \frac{Mv^2}{64} \text{ foot-pounds.}$$

QUESTIONS ON CHAPTER IX

1. Explain fully the meaning of the term *energy*, and in what respects it differs from the meaning of the term *momentum*. Mention bodies which have energy but not momentum.

2. What is meant by potential energy? Describe experiments in which we saw potential energy converted into energy of motion (or *Kinetic* energy, from the Greek *Kineo*, to move). Describe cases in which kinetic energy is transformed into potential?

3. A stone weighing $\frac{1}{2}$ lb. is thrown upon the roof of a house 60 feet high. How many foot-pounds of energy were used in throwing it? What has become of the energy when it reaches the roof and remains there?

4. Would a train consisting of 40 trucks, each weighing 6 tons, and moving 15 miles an hour, or one of 9 carriages, each weighing 4 tons moving 50 miles an hour, do more damage in a collision?

5. A cellar full of coal is said to be a source of much energy. Explain the statement.

6. Much energy was expended more than 3000 years ago in building the pyramids. In what respect can it be maintained that this energy is not lost but is still available?

7. A boulder, weighing 20 tons, rests on the edge of a precipice 100 feet high; how much energy is stored up in it? Is this energy really available? What would become of it if the boulder were to part with it?

8. If you had to treble the energy of a moving body, how much must you increase its velocity?

9. Explain fully what conditions must be fulfilled in order

that a body may possess (I) potential energy, (II) kinetic energy. Can a body have both kinds of energy at the same time?

10. A horse draws a carriage on a level road at a uniform rate of 5 miles an hour. What kind of energy does the carriage possess? Suppose that the carriage were drawn up a hill; what kind of energy would it possess (I) when going up hill? (II) when arrived at the top of the hill? How would you calculate the energy possessed by the carriage in either case?

CHAPTER X

THE PENDULUM

Experiment 1.—Suspend a bullet of lead from a hook by a fine thread; draw it aside and then leave it to itself.

Observe that the bullet will immediately return to the position of stable equilibrium, in which under ordinary circumstances it would remain at rest; but that it does not remain in this position, but moves past it to the opposite side, returns again, and repeats this motion many times. Gravity compels the body to descend, while at the same time another force acts upon it by reason of the cohesion of the thread and the hook to which the thread is attached, and thus the body is constrained to move in a particular manner down to the lowest position possible to it; in fact its path forms a portion of a circle, of which the length of the thread is the radius, while the point of suspension is the centre. Now in passing from the higher point to which we have raised it to the lowest point which it can occupy, the body passes from a state in which it has a certain amount of potential energy, in virtue of its raised position, to a state in which it has kinetic energy in virtue of its motion; at the lowest point it has a certain velocity, and the energy due to this is precisely sufficient to carry the body beyond the position of equilibrium on the other side to a distance from it as great as that at which it started. In lifting the body for the purpose of the experiment, we have done work upon it against gravity; this work is not expended in vain, for the body has for an instant energy of position. When the body is left to itself, this energy, at the lowest point of the swing, becomes energy of motion exactly equal in amount to the work done upon the body by ourselves in raising it; next, since energy cannot be destroyed, the body is capable of raising itself against gravity to precisely the same height on the other side as that to which we lifted it on this side, converting its kinetic energy completely into potential, and these changes are then repeated again and again.

The kind of motion which we observe here is termed *vibratory* or *oscillatory* motion; a body suspended in stable equilibrium, and free to move is called a *pendulum*.

The motion of a pendulum is due to gravity, which continually increases its velocity while it moves from its highest to its lowest point, in the same way as if the body were falling freely through the vertical distance between these two points; similarly while rising from the lowest to the highest point on the other side, the velocity diminishes gradually until it is equal to zero when at the highest point. Thus the motion of a pendulum is very similar to that of a stone thrown upwards, which ascends to a certain point and then falls again. In the latter part of its motion the motion is said to be *accelerated*, in the former it is *retarded*, and in the case of the stone as well as in that of the pendulum, the acceleration of motion at any point of the descent is precisely the same as the retardation at the same point of the ascent. The vibrations of a pendulum would thus always be equal, and its motion would continue for ever, were it not for the effects of the friction at the point of suspension and the resistance of the air, which cause the arcs described by the vibrating body to diminish continually until finally the body is brought to rest; in other words the energy is transferred partly to the surrounding air, which is pushed out of the way, partly to the material particles at and near the point of suspension in the form of heat, and this goes on until the whole of the energy in the moving mass is parted with to other material bodies.

Experiment 2.—Suspend a bullet of lead and one of gypsum side by side as in Fig. 30, by threads of equal

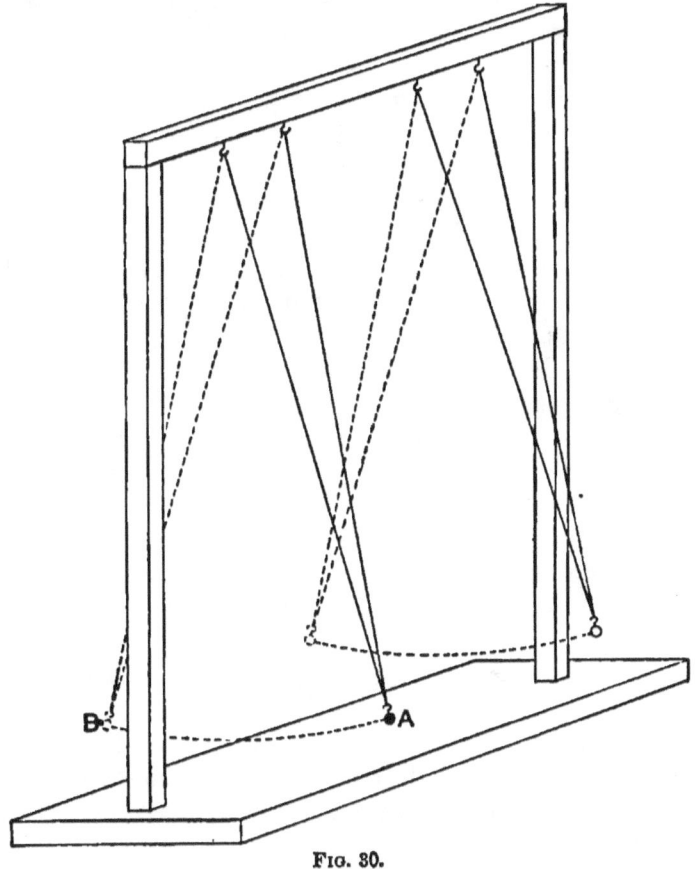

Fig. 30.

length; draw both pendulums back to the same distance from the vertical, and let them go at the same instant.

It will be observed that both pendulums perform their vibrations in the same time, arriving together at the highest point on the opposite side, and again together at the highest point on the side from which they started. It follows from what has been said on the preceding experiment, that both bullets, though they consist of different materials, arrive with the same velocity in the middle of the swing. Now both are acted on by gravity, the descent from the highest to the lowest position being nothing else but a fall from the one point to the other due to the action of gravity; and hence we conclude that gravity acts in the same way upon all bodies, whatever the nature of their substance. The leaden bullet has a larger mass than the other, and is therefore less easily moved, but the action of gravity upon the leaden bullet is greater in precisely the same proportion as that in which its mass is greater, hence the bullet of lead and that of gypsum move with equal velocities.

We should be justified by the preceding experiment in concluding that all bodies fall through equal spaces in equal times, as our two pendulums did; and our conclusion, which is perfectly correct, would be borne out by actual observation, were it not that the resistance of the air acts unequally upon bodies of different shapes and masses, the air presenting to a body of larger surface a greater obstacle during its descent than to a body of equal mass but of smaller surface. Again, if two different masses have the same shape and bulk, the action of the air upon them will be equal, but the action of gravity upon the greater mass will be greater, hence the relative importance of the resistance of the air will be greater in the case of the smaller mass, which will therefore be more retarded than the greater mass.

Experiment 3.—Take a penny and place a disc of paper which is a little smaller than the coin upon it.

Hold the coin in a horizontal position about 2 feet above the table, and drop it.

The paper disc, which, if it were dropped alone, would reach the table much later than the coin, because the resistance of the air would retard it more than the much larger mass of the coin, will in this experiment not lag behind, but will reach the table together with the coin, for the latter pushes the air aside and the paper disc is unresisted in its fall.

Experiment 4.—Repeat Experiment 2, but now use bullets of lead for both the pendulums, and draw one of them farther from the vertical line than the other. Start them both simultaneously.

The arcs through which the two pendulums move, the arcs of vibration as they are called, are now of different lengths, but the pendulum which swings through a larger arc at the beginning of its path moves down a steeper descent than the other, in fact it falls through a larger vertical space, and hence it acquires a greater velocity, which would enable it to perform its vibration in a shorter time were its path of the same length. What we actually observe is that both pendulums arrive at the opposite ends of their arcs at the same instant, in other words, that each pendulum requires the same time for completing each vibration. The name *amplitude* has been given to the magnitude of the arc of vibration of a pendulum, and the above result, which may be stated to be that *the time of vibration of a pendulum is independent of the amplitude*, is sometimes called the law of isochronism (from the Greek *isos*, the same, and *chronos*, time).

Strictly speaking, the above law is only correct when it refers to pendulums which describe very small arcs. Thus, if we were to count the vibrations of two pendulums of which one swings through an arc of 5°, while the other equal to it in all respects swings through 10°, we should find that the latter makes only 998½ vibrations in the time which it takes the other to make 1000 vibrations. This is an important matter, because the laws which regulate the motion of pendulums find a very extended practical application in all branches of Physics and some technical arts.

Experiment 5.—Suspend two pendulums, one of which is precisely four times as long as the other. Draw them aside until each makes exactly the same angle with the vertical, and start them simultaneously.

The short pendulum will be observed to vibrate more quickly than the long one; it arrives at the opposite end of its swing when the long pendulum is in the middle of its arc of vibration, and it has returned to its starting-point, when the longer pendulum arrives at the opposite end of its arc of vibration. While the long pendulum is returning to its starting-point, the short one again completes a whole oscillation to and fro, so that both now arrive simultaneously at their starting-points. Two pendulums which vibrate in unequal times cannot easily be observed together at any intermediate points of their paths, but the simultaneous return of both to the points of starting after each whole oscillation of the long pendulum can be accurately observed, and it may thus be ascertained that the short pendulum makes twice as many vibrations as the long one. If the experiments are repeated with longer pendulums, one pendulum may be made 9 times, and again 16 times as long as the other, and it will be found that in the former case 3, in the latter 4 vibrations of the shorter pendulum are performed in the same time as one vibration of the longer. From these experiments it follows that the time of vibration of a pendulum depends on its length. If a pendulum is to vibrate in twice as long a time as another pendulum, it must be made $2 \times 2 = 4$ times as long, if in three times as long a time, it must be 3 × 3 = 9 times as long, and so on; that is: *the lengths of different pendulums are proportional to the squares of the time of their vibrations.* The time of vibration of a pendulum may hence be varied by varying its length in accordance with this law.

By time of vibration is here meant the time required for a whole oscillation to and fro, that is from the starting-point to the opposite end of the arc and back again to the starting-point, from A in Fig. 30 to B, and back again to A. This is also sometimes called the *complete period* of a pendulum's motion. What is called the time of a single vibration, which is sometimes loosely spoken of as the time of vibration, is that of half of a complete vibration. A pendulum which for such a single vibration, that is for half a complete period, requires precisely one second of time, is called a *seconds pendulum*. The length of the simple seconds pendulum, that is, the distance from the point of suspension of the thread to the centre of the small bullet, is 39·14 inches, that is, this must be its length in London, for its length varies to a small extent in different places on the earth's surface. The reason of this is that the force of gravity, which causes the motion of the pendulum, is different at different parts of the earth, being least at the equator, and greatest at the poles. It follows that a pendulum which is to make one swing in a second must be somewhat shorter at the equator than at the poles, in fact the length of such a pendulum must be increased from 39·01 inches at the equator to 39·22 inches in latitude 80° North.

The pendulum which is used for regulating the motion of a clock consists essentially of a heavy rod and a lens-shaped body, the "bob." In such a *compound* pendulum the time of vibration depends not on the length alone, but also upon the form, size, and relative weight of its component parts. In a simple pendulum the thread is so light that its weight exerts no influence upon the time of vibration, and the bullet is so small that all its points are approximately at the same distance from the point of suspension, and hence move with equal velocities. But in the compound pendulum the rod is of considerable weight; thus, since the portions nearer to the point of suspension would oscillate more rapidly if they were moving independently of the lower portions, but are compelled by the cohesion of the several parts to swing together, the result is that a compound pendulum vibrates more rapidly than a simple pendulum of equal length.

Experiment 6.—Fix a stretched steel wire between two nails, as in Fig. 31, and pluck it with the thumb and forefinger.

The wire, as soon as left to itself, performs a number of vibrations to and fro which are very similar to those of the pendulum. There are, however, several important

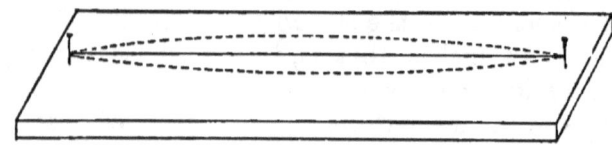

Fig. 31.

points of agreement and difference to be noticed in considering these vibrations and those of a pendulum. In the first place, the work done in plucking the wire is not in this case done against gravity; the wire is lengthened by the work done upon it by us, its parts are somewhat removed from each other in consequence of this lengthening, and hence the work is done against the forces of cohesion and elasticity, which tend to keep the parts of the wire together. In the next place, as regards the energy of such vibrations, the case is very analogous to that of a pendulum vibrating in consequence of the action of gravity. The various particles of the body alternately possess energy due to actual motion when in the middle of a vibration, and energy due to position when reaching the limit of their excursion on each side of the position of rest. The particles vibrate alternately on both sides of that position of rest, and in passing through this position they are always, like the bob of the pendulum, moving fastest; they have then the maximum value of their kinetic energy, while at the moment when they are at the end of the swing their kinetic energy is *nil*, because it has been wholly converted into energy of position.

Finally these vibrations come to an end like those of the pendulum, and since energy is indestructible, we ask what becomes of the energy possessed by our swinging wire. In the case of the pendulum the bob and string are continually setting air in motion; the air particles near them take up energy of motion, which is given up again to other particles, and finally the whole is converted, by friction between them, into heat, just as we have seen the energy of motion of a hammer converted into heat, when the motion of the hammer was stopped by its striking against another piece of iron or lead. Again, the friction of the hook which carries the pendulum against its support partially stops its motion, and we know that friction is a source of heat. Thus we have reason to believe that the whole energy imparted to the pendulum is finally converted into that form of energy which we call heat. Now the same is the ultimate fate of the energy given to our vibrating steel wire, as may be observed, but there is an important difference to be noted in the succession of events. Nothing is heard when a pendulum swings, but here while the vibrating body gives up to the air part of its motion, the motion of the air itself takes place with vibrations of that kind which produce upon our ears the impression called sound, and comparing in our mind the comparative slowness with which even our shorter pendulum was swinging, and the rapidity with which this wire moves to and fro, we shall from our observations at once conclude that the vibrations which cause sound are in their nature much more rapid than those of a pendulum. No doubt in this case also the energy of the vibrating wire is finally completely converted into heat, but not before it gives rise to that kind of undulation of the air which affects the ear as sound.

QUESTIONS ON CHAPTER X

1. The same pendulum is swung on the surface of the ground at the sea level, then in a deep mine, and finally at the top of a high mountain. What differences do you expect to observe in the number of vibrations made in equal times at the three localities? Explain the cause of the differences.

2. Two pendulums which are swung side by side have the same length, but one has a very heavy bob and the other a very light one. Will there be any difference in the number of vibrations made by each in equal times? Give reasons for your reply.

3. A flat ruler has a round hole near one end and is suspended from a smooth round nail which passes through the hole. Near it a pendulum, of the same length as the ruler, consisting of a thread and a bullet, is suspended. Which of these will vibrate more quickly when made to swing? Why?

4 Describe the way in which a pendulum may be made a correct measurer of time. Mention some practical applications of the pendulum. What is meant by a simple and what by a compound pendulum?

5. Compare the vibrations of a stretched string, fixed at both ends, with those of a pendulum, stating in what respects they are similar and in what respects different.

6. How would you proceed to prove by means of a pendulum that the action of gravity is greater in London than in Madras?

7. Show how to explain, by means of the motion of a pendulum, the meaning of the terms "potential energy" and "kinetic energy." When has the pendulum kinetic energy only? When has it potential energy only? When has it both?

8. How long is a simple pendulum which makes two vibrations in a second?

9. What effect has the length of the arc upon the rate of a pendulum? A certain pendulum swings through 4 degrees of arc and makes one vibration in a second. Will it take more or less time for one vibration when swung through 10 degrees?

10. A pendulum clock is found to lose. What change in it would you make to correct it?

11. Two simple pendulums are 4 and 9 feet long respectively. While the short one makes one vibration, how many will the long one make?

12. A pendulum clock which on an autumn day keeps perfectly correct time in London is taken first to Calcutta in summer, and then to Iceland in winter. Will the clock gain or lose in these places? State your reasons for expecting either one or the other result. Have you any grounds for expecting that even in London it will not always remain correct?

CHAPTER XI

SOUND A FORM OF ENERGY

Experiment 1.—Hold a glass tube at a point whose distance from one end is about one-fourth of the length of the tube, strike the middle of the tube with the knuckle of one finger, and listen attentively.

We have already seen that energy possessed by a moving body is converted into heat when the motion is wholly or partially stopped. But in most of the cases in which motion is stopped, as for example when a moving hammer strikes an anvil, or dropping water falls upon a table or into a basin containing water, or moving air passes through a narrow chink, and in thousands of other instances, we perceive that something affects our sense of hearing and we then say that *sound* is produced. Since sound is never produced unless a body, whether solid, liquid, or gaseous, which is in motion, comes in contact with another body which stops or diminishes the motion, we are justified in concluding that just as heat is a form or kind of energy into which we have seen the visible energy of a moving body can be changed or transformed, so sound may be another form into which the energy of a moving body may be changed; also, that heat-energy is perceived by our sense of touch, but sound-energy by the ear; and further, that just as we have come to the conclusion that what is called heat is simply caused by the smallest particles of the heated body having taken up the energy of visible motion, and that thus one kind of motion is changed into another, so in the production of sound there may also be a simple transmission of motion from one body to another, and a change in the *kind* of motion.

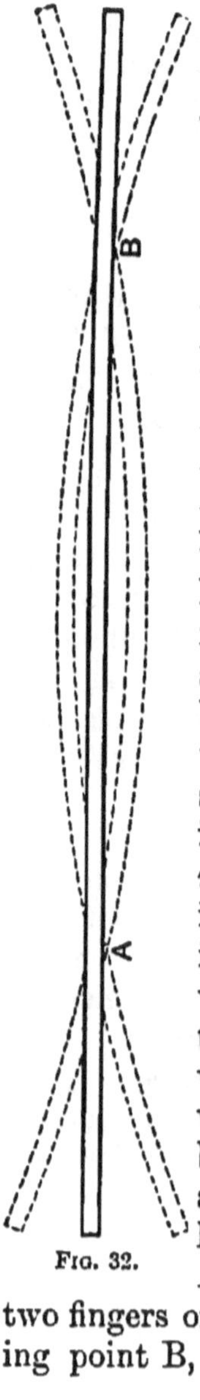

In listening attentively, and as near to the tube as possible, to the sound produced by it, we shall at once recognise that there is an important difference between what we hear in this case, and the sound heard when we strike a table, or a book, or a wall, with the knuckle of the finger. There is at first a sharp sound produced due to the contact of the finger and the tube, and not very unlike that which is heard when a table or book is struck with the finger. But in these cases the sound is short, almost instantaneous, and no motion of the book or the table is perceptible. It is otherwise in our experiment. There is a short sharp sound heard which passes immediately into what is well known to be a *musical tone* or *note*, which lasts during a sensible time, depending chiefly on the force applied to our finger in the first instance, that is, on the amount of visible energy possessed by it when it strikes the tube. If we carefully look at the tube while the note is heard, we shall see that it is vibrating rapidly up and down, not as a whole, but that a portion of it moves upwards while another portion moves downwards, somewhat like the dotted line shown in Fig. 32, where

Fig. 32.

A or B indicates the part held firmly between two fingers of our hand. This point A, and the corresponding point B, whose distance from one end is the same as that of A from the other end, remain at rest while the tube is sounding, all other parts are moving up and down, like a pendulum, only so much more rapidly that close observation is required for studying the motion. We know that a steel bar, an iron ruler, and similar bodies emit musical tones when struck, and by using a *tuning fork* we shall be able to observe the vibratory motion more easily than in the case of the tube.

Experiment 2.—Hold the stem of a tuning fork firmly in your hand, strike the table with the end of one prong and listen.

Here also a sharp short sound is followed by a musical tone of longer duration. A tuning fork is a bent rod of steel, and its mode of vibrating when struck is shown in Fig. 33, the points A and B corresponding to those with the same letters in Fig. 32. We thus see that not only do both the prongs vibrate to and fro, but also the short bent part to which the *stem* is attached vibrates up and down.

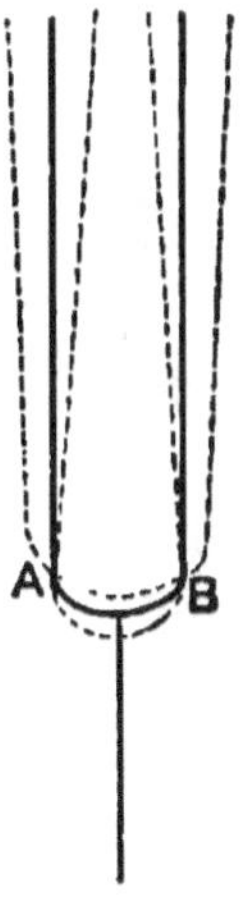

It will not be difficult to understand that a body which vibrates rapidly, like this tuning fork, must produce considerable motion in the particles of the surrounding air. Let us suppose, for the sake of simplicity, that the direction of vibration of one of its prongs is exactly towards and away from us. As soon as the prong commences to move towards us the nearest particles of air are compressed as represented in Fig. 34 at A, a, while particles at b, c, d, are yet unaffected by the impulse. But air being a highly elastic substance, the particles which are being compressed by the advancing prong endeavour to expand again, and this expansion takes place in that direction in which they meet with the least resistance, that is, in the direction towards b where the air is still at its ordinary pressure, while those

FIG. 33.

particles which are near the prong are in a state of greater pressure, and hence of *compression* or *condensation*. We shall now have a state of compression at b, in B, the particles at a moving in the direction of the arrow.

Next let us assume that C represents the prong at the instant when it has just commenced its backward motion, that is, away from us. The particles of air at b will in consequence of their inertia maintain their motion towards c and cause a compression at c, while the particles between a and the prong will expand more and more, they will suffer *rarefaction*, because the prong begins to move to the left, while the particles at a still continue to move to the right. But as soon as this takes place, and the air close to the prong becomes rarefied, and of considerably less pressure than the still somewhat compressed air at b, then the particles a reverse the direction of their motion and move towards the prong, as in D, Fig. 34. The rarefaction is now between a and b while the compression proceeds beyond c towards d. At b the air has now again acquired the original density.

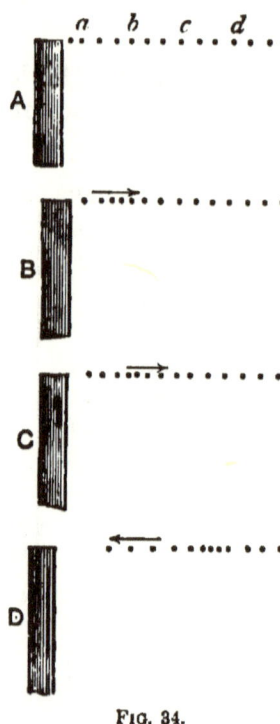

FIG. 34.

In this manner the compression continues to approach us, and is followed by the rarefaction, which advances in a similar manner, because the particles at b now flow towards the rarefied part, causing a rarefaction at b, and so on. Such compressions and rarefactions of the air must take place whenever a body, surrounded by air, is set into rapid vibration, that is, whenever its parts are made to move rapidly to and fro through a short space.

Experiment 3.—Strike the table again with the tuning-fork but immediately place your fingers upon the prongs.

We shall only hear a sharp short sound, but not the musical note by which it was succeeded in the preceding experiment. In other words, the tuning-fork produces a single unmusical sound when simply struck against a hard body, just as a hammer or other body would do when striking a hard surface. In this case we may assume, from what we have learnt so far, that a short series of vibrations consisting of alternating compressions and rarefactions is produced in the surrounding air; this finally reaches the ear, which essentially is nothing else but a portion of the outer covering of the body suitably adapted for receiving such vibratory movements and communicating them to a nerve called the auditory (Latin *audio*, to hear) nerve, which in its turn conveys the impression which it has received to the brain, which is capable of distinguishing the sensations brought to it by the ear from others conveyed to it by the touch, or the eye, or any other of our senses.

Next, we learn from our last experiment that for the production of a musical note it is absolutely necessary that the first comparatively wide and abrupt swing of the prong after striking the table should give way to a continued series of rapid vibrations which follow each other in regular succession. The visible energy of the blow is partly converted into heat, partly into sound, partly into vibratory motion of the prongs which in its turn gives rise to the series of motions of the air which produces the impression of a musical note upon our mind.

Experiment 4.—Sound the tuning-fork as before, and immediately it has been struck hold it in a perpendicular position with the end of the stem touching the table.

The note which we hear will at once become much louder, and we shall actually feel the up and down motion of the stem itself which is shown in Fig. 33 by the dotted line. This vibration of the stem is communicated in some measure to the table, and thus to the surrounding air; we have therefore a much larger solid surface vibrating rapidly and setting a correspondingly larger body of air into vibratory motion, and to this the great increase in the loudness of the note heard is due. The fact of which we have thus convinced ourselves, that a sound which originally is faint may be greatly strengthened by making a larger surface take a share in the vibrations is of immense practical importance. Whenever vibrations are thus communicated by one body to another so that an increase in the sound is produced, we say that *resonance* (Latin, *resonare*, to sound back) is effected. In all stringed instruments, like the piano, violin, etc., reinforcement of sound is necessary, and here the *sounding-board* of thin wood which is placed in intimate connection with the vibrating strings fulfils, but in a more perfect manner, the function of the table in the above experiment. This sounding-board strengthens all the notes which are capable of being sounded by the instrument; and hence the strings of the piano, violin, guitar, etc., owe as much of their loudness to the elastic sounding-board of the instruments as does the fork in our experiment to the table.

Experiment 5.—**Repeat the preceding experiment, holding a watch in your hand, and count the number of seconds during which the note can be heard, first, when the stem is held in the hand, next, when the stem is pressed upon the table.**

Observe that the fainter note heard in the first part of the experiment lasts for a sensibly longer time than the louder note heard when the stem rests on the table. This is easily explained when we consider that in each case the note is originally due to the expenditure of a certain amount of energy on our own part, which as we have stated is partly converted into sound; hence when the sound is louder, the same amount of energy must be sooner exhausted and the sound have a shorter duration, than when the sound is faint.

QUESTIONS ON CHAPTER XI

1. Describe one or two experiments which prove that sound is due to vibrations.

2. Does every vibrating body produce sound? Give instances of bodies which can be made to vibrate, and state what conditions must be fulfilled that sound should be produced by them.

3. In what respects does a musical note differ from an ordinary sound?

4. Describe what is meant when we speak of a *compression* and a *rarefaction* in explaining how sound is produced.

5. Explain fully, and represent by a suitable diagram, the way in which the sound produced when a tuning-fork is struck is carried to the ear.

6. Give as many examples of various modes of producing sound as you can; and state in each case whether the sound thus produced is musical or not.

7. What is meant by *Resonance*?

8. Explain why the sound of a tuning-fork becomes louder when the stem of the fork is pressed upon the table.

CHAPTER XII

WAVES OF SOUND—PITCH—AIR-COLUMNS AS RESONATORS

Experiment 1.—Place a long tube ABC partly filled with smoke upon a table, and set a lighted candle near the narrower end *a* of it (Fig. 35). Strike the table a sharp blow with a book near the wider aperture *b*.

The candle will be instantly put out. Now it is clear that the motion of the book would not extinguish the candle at that distance if the tube were not there; of this we could easily convince ourselves by removing it and striking the table again with the book. It follows that the body of air in the tube serves as a medium, or means, for conveying motion to the candle. But the air in the tube has not left it, for no smoke is seen to issue from the narrow end, hence the motion of the book cannot have caused the air in the tube to move towards the candle, in the same way as the wind or the water of a stream moves. The candle flame was struck by air, and we may say that the disturbance caused by the motion of the book in the air close to it has been transmitted within the tube like a *pulse* until it reached the candle.

It will not be difficult to observe something very similar when a stone is thrown into a pool of water. We say in that case that a *series of waves* is spreading outwards all round the spot where the first disturbance began. But, if we carefully observe the waves produced, or throw a small piece of wood upon the surface, we shall easily convince ourselves that no water really flows towards the margin of the pool, though waves are apparently progressing in rings or circles, which have the centre of disturbance for their common centre; further, the chip of wood will be seen to rise and sink at the place where it happens to be floating, without moving perceptibly towards the margin; moreover, we shall see that the different portions of the disturbed surface are simply performing vibratory movements up and down, and that the wave form of the motion is caused by the fact that each small particle of water commences its motion upward or downward, as the case may be, a little later than the small particle in front of it. Thus while

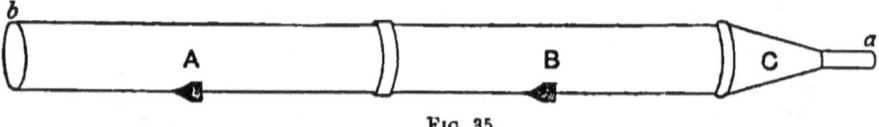

Fig. 35.

one particle has come to the highest point to which the force which produced the disturbance has caused it to rise, and therefore forms the *crest of the wave*, the particle in front of it will at the next instant be also reaching that highest point, but as yet is a little below it; when it actually reaches the top, the crest of the wave has obviously advanced, while the particle which before formed the crest is beginning to sink, but is still as far from the centre of disturbance as before; after a certain time it reaches the lowest point to which it can descend and now forms the *hollow* or *trough of the wave*, whilst some other particle at some distance in front of it is now at its highest point. Between the particle now at the lowest point, and the next particle which is at its highest position there are a number of particles in intermediate conditions as regards their upward and downward movement, and that part of the surface which comprises them constitutes the distance between the hollow and the crest; this is called half a *wave-length*, while the whole wave-length is taken to be the distance between hollow and hollow, or between crest and crest.

Now the motion of the particles of air during the propagation of sound resembles to some extent that of the particles of water during the propagation of a wave; hence sound is said to be propagated by an undulatory (Latin *unda*, a wave) or wave-like motion of particles of air. The resemblance is, however, confined to the fact that each particle, whether of water or of air, performs the same definite movement, but commences its motion somewhat later than the preceding one. But there are essential differences between the two motions. When a wave is propagated in water, each particle moves up and down at right angles to the direction in which the wave progresses; when sound is propagated through air, each particle moves forwards and backwards in the straight line in which the sound is proceeding. Again, waves of water are formed by a series of crests and hollows, or elevations and depressions; sound-waves consist of a series of alternating compressions and rarefactions of air, which are produced in the manner described in the last chapter. In both kinds of wave, however, energy is in the first instance expended in order to produce the vibrations; and as the ultimate effect, whether the margin of the pool is washed by the water-waves, or our ear is struck by the sound-waves, is that work is done at a distance from the source of disturbance, and hence energy is expended by the water and the air respectively without an actual flow of water or air, as in the current of a river or the motion of air in the wind, we conclude that energy may be transmitted by vibrations as truly as by the actual transfer of material from one place to another.

Experiment 2.—Relight the candle and blow through the tube.

The candle will be extinguished as before, but a quantity of smoke will issue from the narrow end at the same time, and prove that the energy of the blowing has set the air in the tube in motion, and that the candle has been blown out as by a wind.

The difference between the propagation of sound through air and the mere motion of a body of air from one place to another can be shown in a very interesting way by an easily constructed apparatus (*vide* "Hints" on the present chapter at the end of the book) exactly resembling a child's drum, of which one head has been replaced by a disc of stout pasteboard with a round hole in its centre.

Experiment 3.—Fill the drum with smoke. Hold it horizontally with the circular hole away from you, and give a single moderately strong tap with your finger upon the elastic cover near you.

Smoke will be expelled through the circular aperture in the form of a beautiful ring, which moves with a velocity which is at first great but afterwards diminishes, the ring itself gradually becoming larger.

It will be at once seen that the smoke ring, although propelled in this case with considerable velocity, travels much more slowly than sound. If the experiment is tried in an ordinary room, some 4 or 5 yards long, an observer stationed at the end remote from the apparatus would not hear the sound of the tapping sensibly later than a person at the other end of the room and close to the apparatus where the sound was produced. The former observer on the other hand sees the rings approach him in a comparatively leisurely manner, and feels their impact if they strike his face, while the impact of sound-waves is not felt unless the sound is of extreme loudness and produced in close proximity, as, for instance, the report of a discharged cannon. The experiment thus proves very conclusively that sound is not due to a motion of translation of air.

In this case also the flame of a candle may be blown out by the moving air at a distance of several yards, if the ring moves so that a portion of its circumference strikes the wick; on the other hand the sound, even if much louder, would have no effect whatever upon the candle flame. Since sound is produced by the motion backwards and forwards of each individual particle of air through a small distance, the objects which the vibrating particles strike are not perceptibly moved; but in the smoke-ring apparatus the whole mass of air which is propelled moves onward in a definite direction, and its collision with bodies in its own path produces an effect which is comparatively great.

Experiment 4.—**Attach a fine wire to one prong of a large tuning-fork (Fig. 36) by means of some soft wax, set the fork in vibration, and quickly draw the point of the wire lightly over a piece of smoked glass.**

A beautiful wavy line will be traced on the glass, each wave corresponding to a vibration of the prong, which thus

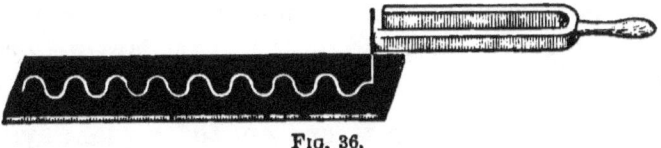

FIG. 36.

traces its own movement while sounding. We have thus a means of recording the vibrations made by our fork.

Experiment 5.—**Repeat the last experiment with one or two other forks, each being somewhat smaller than the preceding one.**

If you do your best to move each fork while it is sounding along the smoked glass at the same rate, employing for example a second, or half a second for the purpose, you will observe that the wavy curves produced by the forks differ essentially from one another in the number of vibrations which they show as having been made by each fork in the same time, and that this number is intimately connected with the character of the tone which we hear while each fork is sounding. The largest fork gives out a note of which we say that it is *lower* than that of the smallest fork, the note of which is said to be *higher* than that of the other; or we may describe this difference by saying that the *pitch* of the larger fork is lower than the pitch of the smaller. Our experiment shows that pitch depends on the number of vibrations made by the fork in a given time, say in a second; the more vibrations per second the higher the pitch, the less vibrations per second the lower the pitch. And it is quite immaterial in what manner and by what instrument a musical note may be produced, whether by a piano, a violin, a flute, or a trumpet, a whistle, an organ-pipe, etc., the law that the pitch of the note depends on the vibrations by which it is fundamentally produced holds in every case.

We shall in Part III. of this course determine, by a more accurate method, the number of vibrations to which a given note is due; for the present our experiment is sufficient to prove the truth of the above law in a general manner.

Experiment 6.—Thrust one end of a glass cylinder A (Fig. 37) into a vessel of water and hold over the other end a vibrating tuning-fork of pitch A′, i.e. making 440 vibrations in a second. Gradually lower the tube into the water until the distance oc is nearly, but not quite 8 inches.

Observe that the sound of the fork suddenly becomes very loud. It thus appears that the column of air in the tube beneath the fork acts like a sounding-board, and

CHAP. XII SOUND-WAVES—PITCH—AIR-COLUMNS 121

reinforces the waves of sound. Instruments which enclose such columns of air are called *resonators*, but it is not the substance of the tube in which the air is contained, but the air itself which strengthens the sound of the fork, and therefore produces resonance.

The following consideration will soon enable us to recognise the cause of the resonance. When the prong a (Fig. 37) vibrates from its position a' to a'', it sends a condensation down the tube, which passes on until it reaches the surface of the water. Here it is thrown back, or *reflected*, just as a billiard ball which strikes perpendicularly against a cushion is thrown back in a perpendicular direction. When this condensation reaches the top o of the tube the condensed air will rush out of the end of the tube so violently as to leave a rarefaction behind. This rarefaction will travel down the tube, and will be reflected at c just as in the case of the condensation, and will travel up again to o. When it reaches this point the air from outside will rush in so violently as to produce another condensation, which will travel along the tube as at first.

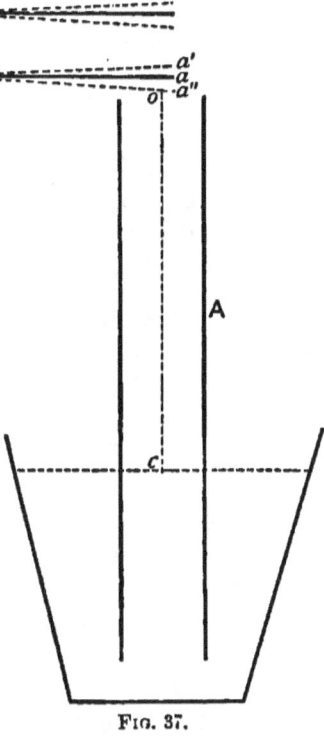

Fig. 37.

Now if the distance oc has a particular magnitude, the first condensation will have returned to o and have been turned into a rarefaction just at the moment when the prong a is returning from a'' to a' and so producing an additional rarefaction. Thus in such a case each rarefaction is intensified, and it is easily seen that each condensation is intensified in the same way, and thus the sound which is heard is increased by resonance.

It must be clear that this can only happen when the distance oc is precisely of such length that the like parts of the waves coincide each with each; if the distance oc were altered, it is plain that condensations would meet rarefactions and thus tend to destroy one another, or in other words to weaken the sound.

Experiment 7.—First push the tube a few inches farther down into the water; next raise the tube a few inches out of the water, and in each position sound the fork as before. Again, sound a number of different forks and observe the length of oc which gives the loudest sound with each fork.

The sound will not only be not strengthened when the length oc is altered, but in particular positions it will be possible to observe that the note is actually fainter than it would be without the tube. Again, the length oc will be found to differ for different forks. We thus see a striking difference between an air-resonator, which can only respond loudly to a definite note, and a sounding-board which responds to a great number.

The experiments just described have another important bearing. They afford us a ready method of measuring the rate at which sound travels from one point in our atmosphere to another. We shall repeat the experiments again in Part III., and there go through the accurate measurements which are requisite for the purpose and explain the steps of our calculation. In the meantime it will be sufficient to state that the velocity of sound in air is as nearly as possible 1100 feet per second.

QUESTIONS ON CHAPTER XII

1. Describe the experiment of blowing out a candle by means of a tube, and state what kind of motion of the air you think takes place within the tube.

2. How are the waves produced which are seen on the surface of water after a stone has been dropped in?

3. How could it be shown that the water is not really flowing towards the margin of a pool or lake or the shore of the sea, when waves are observed on the surface?

4. The candle is blown out in both the Experiments 1 and 2; in what does the difference as regards the transmission of the energy required for blowing it out consist?

5. Explain fully the object of the experiments with the smoke-ring apparatus.

6. Describe how the number of vibrations made by a tuning-fork in a given time may be recorded.

7. Could you suggest a way of representing the number of vibrations made by a tuning-fork in a second, in such a manner that they may be accurately counted?

8. What is meant by the pitch of a note? By what is the difference of pitch of different notes caused?

9. Describe how air may be made to strengthen sound.

10. Mention some musical instruments in which you believe the air to act as a resonator.

11. In what respect does a sounding box differ from an air resonator?

12. Explain why the distance of the top of the tube from the surface of the water in it should be different when we employ tuning-forks of different pitches.

CHAPTER XIII

RADIANT ENERGY

Experiment 1.—Allow the sun's rays to pass through a "burning-glass" and to fall upon a piece of black paper. Move the paper backwards and forwards until the bright speck on the paper is as small as possible. Hold the paper steadily in this position for a little time.

The small portion of the paper on which the rays of the sun appear to be concentrated will soon begin to get very hot and will probably burn. We know by our experience from early childhood that we are warmed when exposed to the sun, that is, that our sense of touch is affected; we also know that we receive light from the sun, and that thereby our sense of sight is affected. The more our knowledge increases the more we shall learn that the sun is the chief source of all the energy which manifests itself under a variety of forms on the earth, in other words, that nearly all the work done on the earth is due to the sun. It follows that the energy of the sun must be brought from it to us in some way or other, in order that it should be rendered capable of doing work where we are. The paper which is used in our experiment possesses potential energy, that is, when raised to a high temperature it is chemically decomposed into various elements, and these in combining with the oxygen of the air around us produce heat and light. Now these elements of the paper are material bodies, they must separate from one another, that is, move away from one another and towards the oxygen, and the oxygen must move towards them, in order to produce the changes which we see by the results must have taken place. This setting in motion of different elements is the work which the sun is actually doing before us at this very moment when the paper gets hot and burns; and as motion cannot take place suddenly of itself, we must conclude that something must be moving between the sun and our burning-glass and paper, thus bringing the sun's energy down to us, and as empty space cannot communicate motion, it must follow that the space between us and the sun is occupied by some substance. This substance, of which we only know for certain that it brings motion from the sun to the earth, has been called *Aether*. We cannot see, hear, feel, taste, or smell it. Nor have we been able to weigh it. Yet without it the indisputable fact that we receive energy, that is, motion from the sun, would be quite unintelligible and inexplicable. The transmission of energy through the medium of the aether is called *radiation*; energy so transmitted is called *radiant energy*, and the body transmitting energy in this manner is called a *radiator*.

It will not be difficult, after having learnt how, by means of waves in air and water, energy is carried along from one point to another at some great distance by vibrations, while no matter is transferred at the same time to a distance, to understand that the sun's energy really reaches us by vibrations of the aether, and that the material particles of a body which is heated by the sun are really set in motion by the vibrations of the aether. Some at least of these vibrations of the aether are capable of causing the sensation of light by affecting the eye. Others again, as we shall soon see, are able to promote in various substances certain chemical actions, which differ in many respects from the mere burning of a substance.

Having now obtained a clearer insight into the manner in which the energy from the distant sun is enabled to cause our paper to burn, we shall be prepared to study the details of our experiment more closely, but we shall first extend the experiment to the case of a different substance.

Experiment 2.—Repeat Experiment 1, using a piece of tinfoil blackened on one side instead of paper. Expose first the bright side, and then the blackened side of the foil to the rays of the sun.

No perceptible effect is produced when the bright side of the tinfoil is exposed, but the foil immediately melts at the spot where the rays are concentrated, when the blackened side is turned to the sun.

We know from ordinary observation that when the sun's rays pass through a window pane, a bright patch of light is

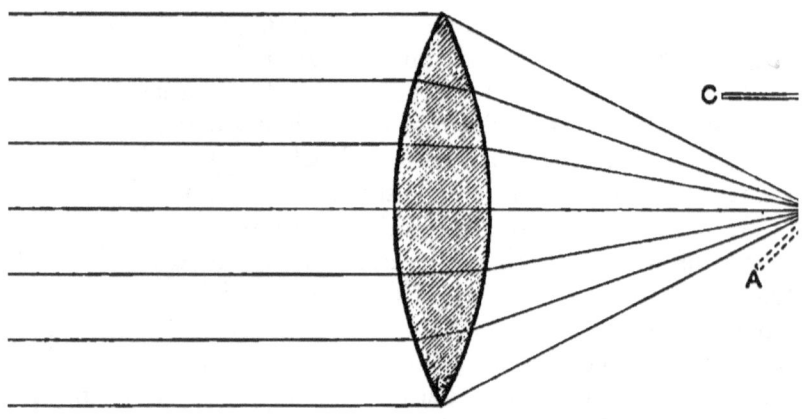

FIG. 38.

seen on the opposite wall or floor of the room, surrounded by the dark shadow of the sash or frame. The light of the sun seems to pass in a straight line through the window and to the picture or *image* of the window on the wall or floor. In our experiment the path which the sun's rays pursue after passing through the burning glass, or *lens*, as it is called, is clearly not the same straight line as that in which they reach the lens; all the rays, except those through the central portion of the lens, appear to be bent from their original direction by the action of the lens itself, and are made to unite or *converge* (Latin, *convergere*, to incline together), as shown in Fig. 38, within a comparatively very small space, which in consequence of receiving all the heat and light of the rays which fall on the whole surface of the lens becomes both very bright and very hot, as is shown by the result. The small space F, where our paper ignites or the tin melts is called the *focus* (Latin, = a burning place, hearth) of the lens, while the bending of the rays from one direction to another effected by their passage through any substance is called *refraction* (Latin, *refrangere*, to break back) of light, or heat, for our experiment clearly proves that the rays of heat are refracted by the lens as well as the rays of light.

When the bright surface of the tinfoil is presented to the rays which have passed through the lens, the foil will not melt, as is the case when the blackened side is held at the focus. The rays apparently suffer some further change which deprives them of their heating effect when they reach the polished bright metallic surface.

Experiment 3.—Repeat the experiment with the tinfoil, but hold it in an inclined position as AB in Fig. 38. First expose the bright side and then the dark side, and place some dark object, such as a book, near the focus in the position CD. Again, hold the back of your hand near the tinfoil in the same position CD.

Observe that rays of light and heat are thrown upon the object CD when the bright side of the tinfoil is in the position AB; but that while the tinfoil will melt when the dark side is exposed at AB, no heat or light is observable at CD in that case. We thus see that in the former case the rays of light and heat must be again bent away, and therefore prevented from producing much light and heat at the focus. This change of direction is, however, different from that effected by the lens, for the lens acts while the rays are passing through it, while the tinfoil throws the rays back as soon as they reach its surface. This throwing back of light and heat waves is called reflection (Latin, *reflectere*, to bend back). We know from common experience that something very similar happens to waves of sound, when sound is produced at some distance from a wall or other obstacle; we hear the sound which is thrown back, and call this kind of reflection of sound-waves an *echo*. In the case of the darkened side of the tinfoil, where no appreciable reflection takes place, and also in the case of the black paper, we say that the heat and light are *absorbed* (Latin, *absorbere*, to suck in), and it is obvious that bodies which are good reflectors like the bright tinfoil must be bad absorbers; while bodies which readily absorb the waves of heat or light which reach them are bad reflectors, as is the case with the substance which forms the black surface of the tinfoil, and which is really impure carbon, or lampblack.

In general we may therefore conclude from our experiment that radiant energy from the sun may meet with a fate which is different according to the nature of the body which is exposed to its action; it may be *transmitted* through the substance of the body, or it may be *reflected*, or it may be *absorbed*, and the path of the waves may suffer some change of direction, respectively called *refraction* and *reflection* in their transmission as well as in their reflection, according as they are transmitted through the body or thrown back from it.

Experiment 4.—Boil some water in a flat, bright, tin canister or flask, of which one side has been previously blackened. Hold the backs of your two

hands about half an inch from the canister, one towards the black, and the other towards the bright side.

The hand opposite to the blackened side will feel the heat much more than the other hand. The blackened surface is at the same temperature as the bright surface, but it radiates much more heat than the bright surface. The same fact may be noticed in the case of all other bodies. Substances which have the greatest absorbing power for heat have also the greatest radiating or *emissive* (Latin, *emittere*, to send out) power. On the other hand, those substances which have the greatest reflecting power, and hence little absorbing power, also possess the least emissive power.

Experiment 5.—Take two discs of the same metal, one of which has a polished surface and the other a rough surface, and expose them side by side to the rays of the sun; leave them there together for six or seven minutes, or longer if necessary, and then take them up again.

The disc with the rough side will feel much warmer than the other, which will be scarcely warmed at all. Thus rough surfaces absorb heat better than polished ones.

Experiment 6.—Make some chalk marks near the end of a poker or piece of iron gas-piping, and thrust the marked end into a fire. When red-hot take it out and examine it in a room which is as dark as possible.

The chalk marks will appear much darker than that portion of the red hot poker which had no marks on it. We know that the white chalk reflects nearly all the light that falls upon it, while the dark surface of the poker reflects very little, that is, it absorbs most of the light. But in the present state, while red hot, the chalk appears dark because it radiates little light; the rest of the poker appears bright because it emits more light. Thus the same relation holds good for light as for heat; surfaces which reflect well are bad absorbers and bad radiators; surfaces which reflect badly are good absorbers and good radiators.

Experiment 7.—Heat a piece of platinum foil, upon which some letters have been written with ink, over a Bunsen burner in a darkened room, first, with the writing turned upwards, next, with the writing turned downwards.

Observe that in the first case the letters appear brighter when the foil is red hot than the remainder of the foil. This result is in accordance with that of the preceding experiment; under ordinary circumstances the black and rough surface of the letters reflects little and hence absorbs more light than the bright surface of the foil, which is highly polished, and thus reflects considerably while it absorbs and hence radiates very little light, and hence when the foil is heated to redness the letters radiate more heat and light than the polished portion. But when the plain side of the foil is turned upwards and the letters downwards, the letters, though they appear inverted as in a looking-glass, are quite legible and appear dark on a bright ground. The reason is that there is more radiation of *heat* taking place on the opposite side of the foil from the lettered portion than from the remainder, including that side at which we are looking, and that hence the metal is cooler just where the letters are written than elsewhere; now the radiation all over the plane surface which is turned upwards would be everywhere the same if it were at the same temperature, because the surface is uniform; but where this surface is colder, as just over the different letters, it cannot radiate so much as where it is hotter; hence on the side at which we are now looking the parts over the letters appear darker than the remainder of the foil. Thus the present experiment confirms the laws derived from the experiments that precede it, and these laws as we see apply alike to the case of heat and to that of light.

Experiment 8.—Set up the air thermometer (see Part I, page 42), and bring the bulb into the focus of a burning glass exposed to the sun's rays.

Note that the radiation concentrated on the enclosed air scarcely affects the instrument, although it is very delicate, and although the temperature at the focus is high enough to ignite paper and to melt tin.

Experiment 9.—Cover the outside of the bulb of the air thermometer with lamp-black and repeat the preceding experiment.

The liquid will be seen to be pushed down rapidly by the expanding air, hence the air must have been heated. In this case the lamp-black absorbs the radiant heat, and the heat being conducted through the glass to the air in the bulb, the temperature of the air is raised. We may then ask, what becomes of the heat in the previous experiment? It is clearly a very similar case to that of an ordinary window pane, which remains quite cold while the sun's rays pass through it and warm objects in the room considerably. The radiant heat which passes into the bulb on one side passes out again on the other without perceptibly warming the glass itself or the air enclosed. Thus radiant energy is not only emitted, absorbed, and reflected by different bodies in different degrees, but it is also able, whether as light or heat, actually to pass through the substance of a body; and different bodies permit the passage of radiant heat not only in different degrees, but even exert a selection according to the source or rather kind of heat which, as it were, asks for permission to pass through them. Thus the heat-waves from a body which gives out light at the same time, such as a white hot poker, or the sun itself, may be called *luminous heat*, while the heat-waves reaching us from a vessel containing hot water, or any other hot but not luminous body, or the heat reflected by bodies which have received luminous heat from the sun, as for instance the ground, with its plants and other objects upon it, may be called *dark* or *non-luminous heat*. Now our atmosphere allows both kinds of heat to pass through it and hence it is *diathermanous* (Greek, *dia*, through, and *therme*, heat) ; but the vapour of water which is present in the air in variable quantities is diathermanous only for the luminous heat of the sun, but comparatively *athermanous* (Greek, *a*, not) to the dark heat reflected from the ground. Similarly glass does not screen us from the sun's heat, that is, it is diathermanous to the sun's radiation, but quite athermanous to the radiation of non-luminous heat. This is well seen in the case of hot-beds and green-houses. The sun's heat passes through the glass of these enclosures almost unobstructed, and heats the earth ; but the radiation given out in turn by the earth is all dark heat which cannot pass out through the glass, hence it is retained within the enclosures.

QUESTIONS ON CHAPTER XIII

1. Explain why in Experiment 1 black and not white paper is used. What difference would it make in the result if white paper were used? (Try the experiment.)

2. Point out any similarities and dissimilarities between the propagation of sound and that of light which the experiments in the last chapter call to your mind.

3. Make a sketch of the appearance presented on a sheet of paper held half-way between the focus and the lens (Fig. 38) in making Experiment 1, and explain (I) why the circular patch of light seen on the paper is brighter along the circumference than towards the middle; (II) Why the middle is less luminous than the part of the paper which is beyond the action of the lens altogether and is simply exposed to the sunlight.

4. Describe the way in which it is shown that a good reflector of heat is a bad absorber.

5. How was the fact that a good absorber is a good radiator demonstrated?

6. What kinds of surfaces are in general good and bad absorbers respectively? How has this been shown by experiment?

7. How has it been shown that the laws of heat radiation also apply to light?

8. Give an account of the experiment made with a piece of platinum foil on which some letters are written with ink, and explain fully the conclusions that can be drawn from it.

9. Rock-salt is the most diathermanous of all solid bodies. What would be the difference between the effects produced upon two thermometers, if one of them is placed behind a plate of rock-

salt, and the other behind a plate of glass, and both plates are exposed (I) to the rays from the sun; (II) to the heat radiated from a kettle of hot water?

10. The presence of vapour of water in the air is said to serve as a "blanket" to the earth. What facts would you mention in proof and explanation of this statement?

11. If you wish to keep a liquid hot as long as possible, would you put it into a vessel whose outer surface was polished and bright, or into one whose outer surface was dark?

12. Explain why we usually wear dark clothes in winter, and bright or light coloured clothes in summer.

13. A black and a white piece of cloth are placed on the snow while the sun is shining on a winter's day. One cloth will sink into the snow, the other will remain on the surface. Which one will sink? Why? (Try the experiment if possible.)

14. Why does snow melt sooner near the trunks of trees than elsewhere?

CHAPTER XIV

REFLECTION

Experiment 1.—Draw two lines at right angles to each other on the table (Fig. 39), and place a heavy block A so that one edge coincides with one of the lines. Place two blocks B and C edgewise as shown in the figure

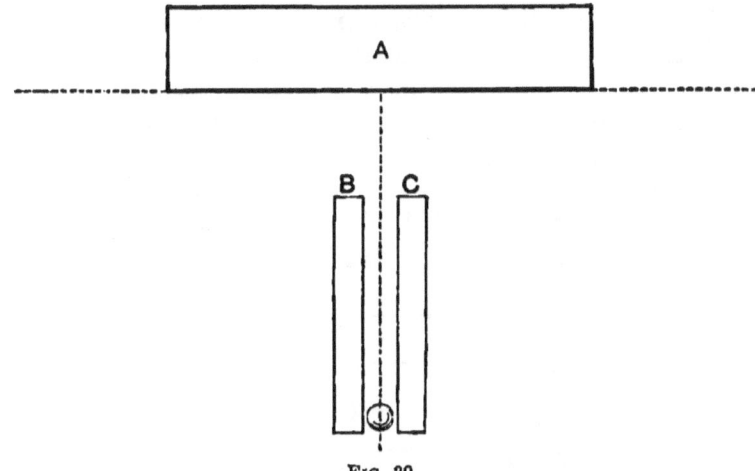

Fig. 39.

parallel to the line perpendicular to the block A, and gently project a small india-rubber ball between B and C so as to strike A.

When the ball comes into contact with the wooden block each of them suffers a slight compression, but both possessing a certain amount of elasticity they recover their original form, and this recovery, due to elasticity, causes the ball to be thrown back, or reflected, in a perpendicular direction, that is, along the same line as that in which it was projected. There is, indeed, no reason why it should take any other path after reflection. The motion during reflection takes place in the same manner as if a force had acted upon the ball in a direction exactly opposite to that by which it was originally set in motion. The elastic force has been called into play in a direction perpendicular to the block, and hence the motion which it produced takes place in that direction.

It follows from what we here see in the case of an elastic ball that when such highly elastic substances as air and aether, moving so as to form waves of sound, and of

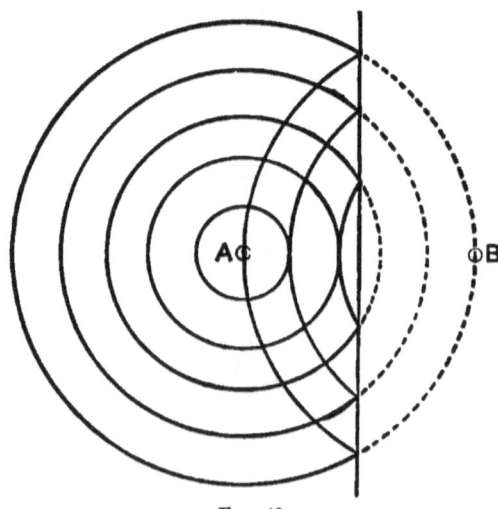

Fig. 40.

heat or light, strike a smooth surface in a perpendicular direction, and reflection takes place, the waves turn back in a direction perpendicular to the surface which stopped their progress.

It will not be difficult to observe the reflection of waves of water at the margin of a sheet of water, especially when some obstacle, such as a wall, presents a straight line to the "front" of a wave, that is, to the direction in which the wave progresses. If the surface is circular, and we suppose the disturbance which gave rise to the formation of a wave to have happened in the centre A, in Fig. 40, the wave is seen gradually to enlarge and to spread in the form of a ring until it reaches the side of the enclosure; here the wave does not disappear, but it returns again, forming a ring which becomes gradually narrower, and contracts at the middle. This is the "reflected" wave. If, however, the reflection takes place at a straight wall, as shown on the right in Fig. 40, the reflected wave does not return, but spreads out backwards in the same manner as if it had been produced at a point B as far behind the wall as the point where it was really produced is in front of it.

Experiment 2.—Draw a line AO (Fig. 41) on the table,

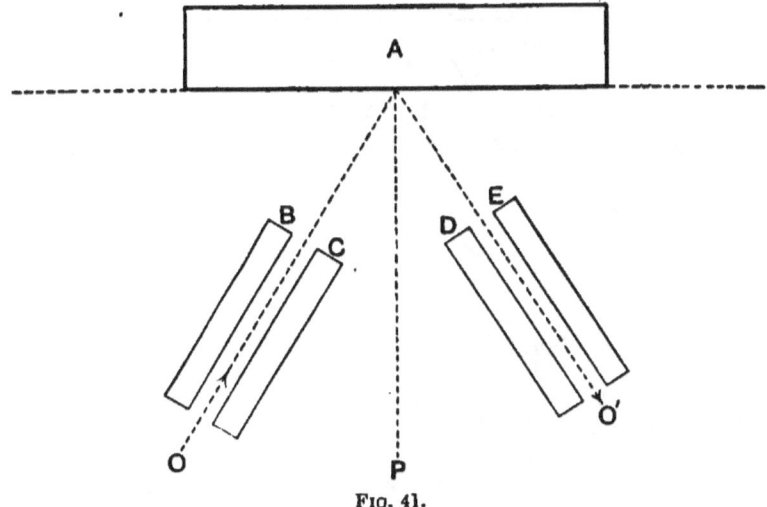

Fig. 41.

making an angle OAP with the perpendicular AP; adjust

two blocks B and C as in the figure, and project the ball between them toward A. Note the direction AO' along which the ball rebounds, and adjust two blocks D and E on each side of this line. Repeat the experiment until the ball in rebounding passes exactly between D and E. Measure the angle PAO'.

The direction OA in which the ball is projected makes with the line AP, which is perpendicular to the reflecting surface at the point A, an angle OAP, which is called the *angle of incidence* (Latin, *incidere*, to fall upon); again the line AO' along which the ball returns makes with the same perpendicular (or "normal," as it is sometimes called) the angle PAO', which is called the *angle of reflection*. Now, as

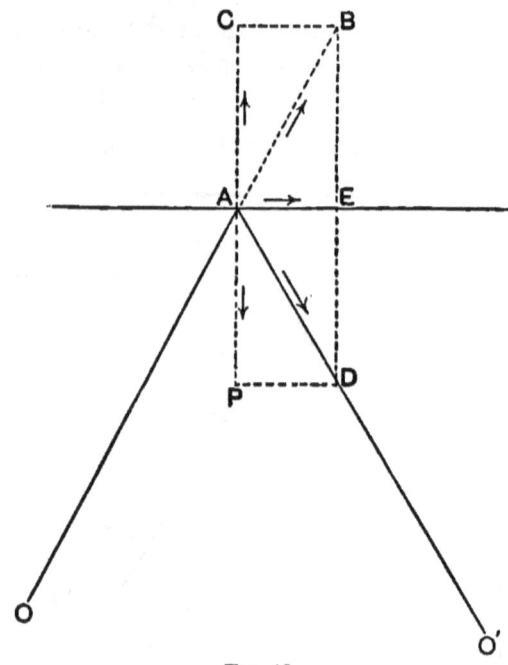

Fig. 42.

the result of our experiment, we shall find that the angle

OAP is, as nearly as the nature of the experiment permits, equal to the angle PAO'.

Experiment 3.—Repeat the experiment twice or three times, taking the angle OAP each time of a different magnitude, and adjusting the blocks D and E accordingly.

Observe as the general result of all experiments that *the angle of reflection is equal to the angle of incidence.* This result, which is called the *law of reflection*, is an immediate consequence of the mode in which all forces act. We may consider the magnitude of the force applied to the ball as represented by the line AB, Fig. 42, drawn in the direction in which the ball is propelled. But a force represented by AB, from what has been stated previously, is equivalent to two forces, one along the face of the block and the other perpendicular to it, represented by the two sides of the parallelogram AEBC, viz. AE and AC. Again AC calls into play a force AP opposite and equal to it, while AE still remains in action. The forces which cause the rebound thus act in the directions AE and AP, and give as their resultant AD, a force in the direction AO' in which the ball will therefore proceed after the rebound. Now it is easily perceived that the figure AEBC is in all respects equal to the figure AEDP, hence the angle DAP is equal to the angle CAB, which angle is equal to the vertical and opposite angle OAP. It follows that OAP, the angle of incidence, must be equal to the angle PAO', the angle of reflection.

We must not, however, forget that the energy which the projected ball has at the moment of impact with the block is in reality partly converted into heat, partly into sound, and that the effect of this is seen in the rebound, which will not proceed exactly in accordance with the above assumptions. When the ball returns it has lost a certain amount of energy, that is a certain amount of velocity, hence on its return the effect is the same as if a less force had set it in motion; thus AP in the figure will not be exactly equal to AC, and the angle of reflection not exactly equal to the angle of incidence. But it is different in the case of light, and hence the law can be demonstrated to be perfectly true by the most refined astronomical observations. For our purpose we may proceed to the demonstration with an apparatus which is very easily constructed, as is explained in the Appendix of "Hints." It is shown in Fig. 43. The apparatus consists of a shallow

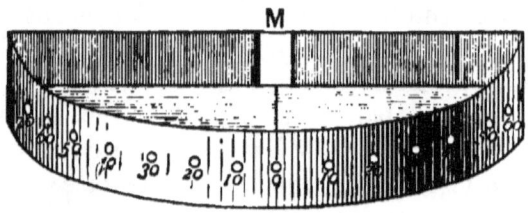

Fig. 43.

semicircular vessel. The curved side turned towards the observer is divided into 18 equal parts, that is, at intervals of 10°. The division in the middle is marked 0, the two on each side next to it are marked 10, the next two 20, and so on as far as 90. Beginning with 0° there are apertures at every 10°. Opposite to the aperture at 0, in the flat side of the vessel, is a small mirror M, which is so adjusted that on looking through the aperture opposite to it the reflected image of the aperture is seen exactly in the middle of the mirror, which is only the case when the straight line drawn from the aperture to the mirror is perpendicular to the mirror, just as the ball, in our previous experiment, will only rebound in the same straight line if it is projected along a perpendicular to the reflecting surface.

Experiment 4.—Look through the hole marked zero at the mirror. Then place a candle flame before it and look through the other holes.

You can see in the mirror an image of the hole marked zero through which you are looking, but you do not see the image of any of the other holes. Nor can you see through any of the other holes the image of the candle flame placed before the hole marked 0.

Experiment 5.—Hold a candle flame at the hole marked 10.

You can see the reflected rays of the candle flame only through the hole similarly marked on the other side. As before, nothing is seen through either the hole marked 0, nor through any of the others. The rays of light which make an angle of 10 degrees with the perpendicular at the point of incidence (marked in the figure), that is, having an angle of incidence of 10 degrees, are seen to be reflected so as to make an angle of reflection of 10 degrees on the other side of the perpendicular.

Experiment 6.—Place the candle flame in succession close to the holes marked 20, 30, etc., and apply the eye to the corresponding holes on the other side.

The experiments prove that light is reflected so that

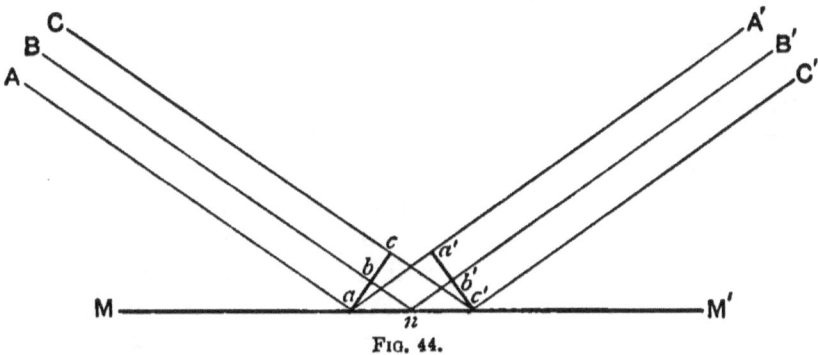

Fig. 44.

the ray of light after reflection makes the same angle with the perpendicular as that made before reflection.

The reflection of light differs materially from the rebound of a solid body like the ball in the former experiment, for we have not to deal with the motion of a body from one point in space to another, but with the transmission of energy by means of wave or undulatory motion. Now suppose A, B, C, in Fig. 44 to represent rays of light parallel to each other, and they transmit the light in the form of aether-waves. Let us also suppose that the points a, b, c, are all forming the "front" of a wave, that is, that they are all in the same "phase" of the vibratory movement, which means that they are forming together either the crest or the hollow of a wave. The line joining them, abc, will thus be perpendicular to the direction in which the wave proceeds, this direction being in the present case oblique to the reflecting surface MM'. When the wave proceeds without any disturbance the front of the wave will in any position be parallel to each previous position which it occupied, but this will no longer hold good when the reflecting surface MM' effects a change in the direction in which the wave moves. On the other hand, the rate at which the various points of the front are travelling remains the same, for no cause exists which would involve a loss of energy in this case; hence the new front of the wave, $a'b'c'$, is such that the disturbance at a, b, c, will have travelled equal distances in order to come into the new positions a', b', c'. Thus the path aa', travelled by a, is equal both to the path bnb' travelled by b and to the path cc' travelled by c. From this it follows immediately that the triangles acc' and $aa'c'$ are in all respects equal (the angles at c and a' being right angles), and hence the angle $a'ac$, which by geometry equals the angle $C'c'M'$ is equal to $cc'a$. From this we see at once, if we conceive a perpendicular erected at the point c' that the angles of incidence and of reflection are equal.

Experiment 7.—Hold an open book with one hand so

that its back is turned towards the window and the pages are in a deep shadow; with the other hand hold a piece of white paper a little farther away and higher than the book.

Observe that the print will appear sufficiently brightly illuminated, at any rate upon the upper portion of the page, to be read easily. Even holding up the hand in the place of the paper will sensibly increase the light which falls upon the page.

Experiment 8.—Repeat the preceding experiment, using a small square piece of looking-glass instead of the paper.

A portion of the page will now appear much brighter than when the paper was held before it; moreover, the portion illuminated will now have a sharply defined boundary separating it from the rest which remains dark. Thus a well-polished surface like that of a mirror reflects light in a definite direction, while rough and dull surfaces

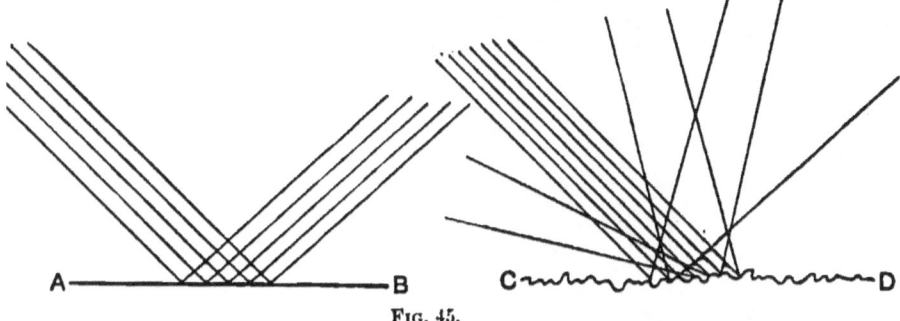

Fig. 45.

reflect light in all directions. Such surfaces scatter or *diffuse* (Latin, *diffundere*, to pour in all directions) the light. The difference in the effects produced is caused by the difference in the smoothness of the two reflecting surfaces. AB in Fig. 45 represents a smooth surface, like that of glass which reflects in the same direction nearly all of a set of parallel rays of light, because nearly all the points of reflection are in one and the same plane. CD represents a surface of paper, having the roughness of its surface greatly exaggerated. The various points of reflection are turned in every possible direction, consequently light is reflected in every direction. It is thus that the surfaces of most objects around us reflect light in all directions, and are consequently visible from every side in consequence of the reflected rays acting upon our organ of sight, the eye. By means of regularly reflected light we see "images" of objects in mirrors, but only in definite directions; by means of diffused or irregularly reflected light we see the reflecting surface itself from whatever direction we look at it. We must, however, not forget that light, although irregularly reflected, still in strictness obeys the law of reflection as proved in this chapter both by direct experiments and conclusions from experiments.

QUESTIONS ON CHAPTER XIV

1. A very long tube is placed so as to point to a distant wall, being inclined to the direction of the wall at an angle of 30 degrees. Show by a diagram where an observer must place himself to hear by reflection the sound of a pistol fired at the mouth of the tube.

2. A and B are two billiard balls placed in a straight line which is parallel to one of the cushions. Show by a diagram how a player must strike A so that it may hit B after rebounding from one of the cushions.

3. Enunciate the law of reflection of light and describe some experimental way of proving it.

4. Describe some way of establishing the same law as a consequence of the nature of wave motion.

5. Two mirrors are inclined at an angle of 60 degrees. A ray of light falls upon one mirror in a direction parallel to the other mirror. Trace accurately its path after being reflected by both mirrors.

6. Explain what is meant by irregular reflection, and draw two diagrams to show how it differs from regular reflection. Is there any real exception from the general law of reflection in this case?

7. How would you explain the fact that on looking at a sheet of water when the sun is nearly setting the reflection of the sun's light is much more brilliant than at noon when the sun is overhead?

8. By looking at a plate-glass window on a sunny day, a person in the street can see passers-by apparently walking inside the house. Explain how this is possible. Why is it noticed on a sunny day only?

CHAPTER XV

REFRACTION

Experiment 1.—Arrange a candle at such a distance from the rectangular vessel A (Fig. 46) that the bottom of the vessel, BD, may be just in the shadow of one of the sides. Then fill the vessel with water.

Observe that the bottom of the vessel, as shown in the figure, is now no longer entirely in the shadow. The rays of light which strike on the surface of the water proceed in the liquid in a more slanting direction than that in which they were propagated in the air. The extreme ray,

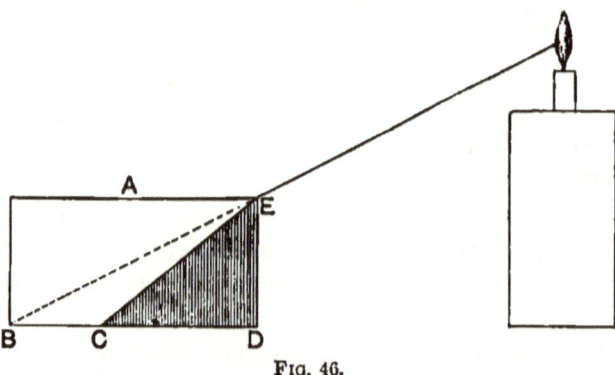

Fig. 46.

which reaches the liquid at the point E, is bent towards the perpendicular side ED of the vessel, hence more rays travel towards the bottom of the vessel; the effect is the same as if the candle had been brought nearer to the vessel, or raised up at its present distance from it. The remarkable fact is thus manifested that rays of light, which we know to proceed in straight lines in air, or in water, or within any other transparent substance, appear to deviate from their path as soon as they leave one substance for another.

Experiment 2.—**Place a piece of wire vertically in front of the eye, and hold a narrow strip of thick plate glass horizontally so as to be across and close to the wire, as shown in Fig. 47, A.**

Attentive observation will reveal the fact that in looking at the wire through the glass it will appear to be broken at the two edges of the glass, and the intervening portion will appear to be displaced towards the left, as in the figure, if we hold the glass and the wire so that we are looking at the glass from a point on the right-hand side of a line drawn perpendicularly to its surface; on the other hand the displacement will be to the right, if we look at the glass and wire from the left side of the perpendicular. Finally, if we hold the glass and wire straight before us no apparent displacement of the wire is observed. We may thus conclude that the ray WE (Fig. 47, B) which proceeds from any point in the wire perpendicularly through the glass suffers

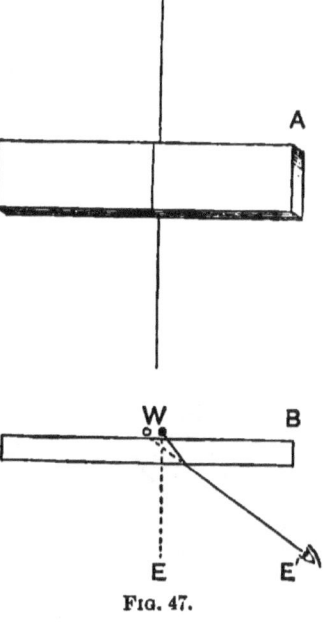

Fig. 47.

no alteration in its direction when leaving the glass and reaching the surface of the air; on the other hand, any other ray which passes through the glass in any other direction but the perpendicular becomes still more slanting when leaving the glass and entering the air, so that an eye, if placed so that this particular ray may reach it after it has been deviated by the glass, sees the object in the direction in which the ray enters it, and hence to the left of the object when the eye is to the right, as is supposed in the figure; and to the right, if the eye is on the left side of the wire.

Experiment 3.—Make a small hole in a piece of cardboard. Fix the cardboard NO (Fig. 48) in a vertical position, and adjust the empty basin ABCD so that a coin near A is just seen when looking through the hole over the edge D. Push the coin towards you to the point F so that it is just out of sight. Then without moving either the card, the coin, or the basin, fill the latter with water, and again look through the aperture.

The coin will now be distinctly seen. A ray of light ADHM passed at first from the coin at A past D to the aperture H. When the coin is displaced the ray FD passes to the point O above the aperture, hence the coin is invisible. But when water is poured into the vessel, in consequence of the deviation produced at the boundary between water and air, the path of the ray is represented by the broken line FDH, hence the eye sees the coin along the

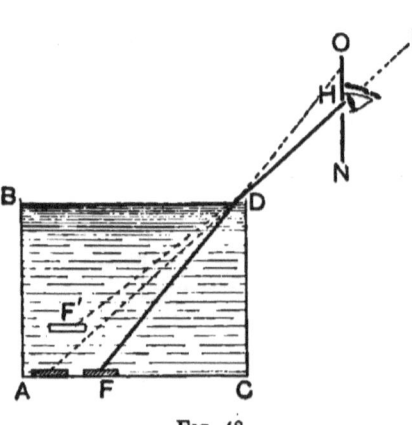

Fig. 48.

direction HD, that is at F', a point which is clearly *above* F. But the coin is still resting on the bottom of the vessel; hence the effect of this bending of the rays of light as they pass obliquely out of the water is to cause the bottom to appear higher up than it really is, or to cause the water to appear shallower than it is.

We also notice that a ray emerging from water into air is bent away from the perpendicular CD, that is it makes a greater angle with this perpendicular outside of the water than in it.

Experiment 4.—Thrust a pencil obliquely into a glass of water.

Notice previously what length of the pencil will be immersed and you will see in a very striking manner that not only does the portion in the water appear bent away from the portion outside, but also the whole pencil seems shortened, and the end of the immersed part appears to be nearer to the top of the liquid than it really is.

From all these experiments it is perfectly obvious that the bending of the rays of light must be due to the fact that the waves of light change their front whenever they pass obliquely from one substance into another. We may inquire into the cause of this change by imitating in a very rough way the conditions of the rays of light in moving from one substance to another.

Experiment 5.— Place one end of a well planed board, a portion of which, having the shape of the trapezium SEFT (Fig. 49), has been covered with emery paper, upon a block, and allow a pencil AB to roll down the inclined plane.

The pencil in rolling down will at the instant it reaches the position ab touch the rougher surface at b, and if this surface were not presenting a slanting edge to the pencil but an edge at right angles to the direction of motion the pencil would continue to move in the same direction as before; its rate of motion, however, would be slower on the rough surface of the emery paper than it has been on the smoother surface of the board, in consequence of the greater friction. Now, in the present case the end b begins to move over the rough surface while a is still moving over a smooth one; as a consequence, while the

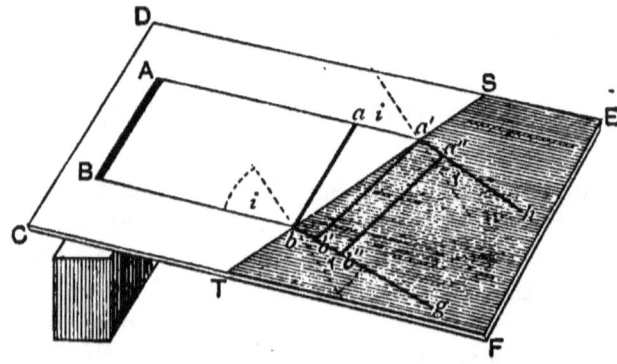

FIG. 49.

end a of the pencil has moved to a', b will have moved through a shorter distance, perhaps to b', and $a'b'$ will be the position of the pencil when every point of it is upon the surface which presents greater resistance to the motion, and therefore retards the velocity; the next position will be at $a''b''$, and so on. In other words, taking the motion only of the two ends of the pencil into consideration, the direction of motion for each of them are the broken lines Bbg, A$a'h$, respectively, the two lines being of course parallel to one another. Now, it is obvious that the motion of the point B, for instance, along the line Bbg is exactly similar to that of a ray of light passing from air into water; this is bent at the surface of the water, and its direction is afterwards nearer to the perpendicular at the point of incidence than it was before, or the angle i is greater than r. It follows that if we consider AB to be the front of a light-wave moving in air, it must change its front on reaching the surface of a transparent substance like water or glass in a slanting direction, if these substances exert a retarding action on the rate of motion of the light-waves. Now, this very fact has been proved by exceedingly laborious and complicated experiments, which cannot easily be made in an ordinary laboratory. Light moves in aether or in air at the rate of nearly 186,000 miles in a second, but it moves through a less space per second in water, and still less in glass.

Thus the experiment just made has partly explained the cause of the bending of an oblique ray of light when it crosses the boundary which separates two transparent substances. The name *refraction* (Latin, *refrangere*, to break again) is given to this bending of a ray. The angle made by the incident ray with the perpendicular to the surface at the point of incidence is called the *angle of incidence*, that between the same perpendicular continued into the other substance and the refracted ray is called the *angle of refraction* (i, r, in Fig. 49). We have already seen that in the case of refraction these two angles are not equal as in that of reflection. The ray of light is always nearer to the perpendicular in the substance or "medium," as it is called, which is denser, which in this case means that substance in which the velocity of light is less. There is, however, a definite relation between these two angles as regards the same two substances—for instance, air and water, or air and glass. This relation will be investigated by some definite measurements, in Part III., for air and glass. In the meantime we may consider the following case: In Fig. 50 the vessel is supposed to contain water, and a ray of light travelling in the air is supposed to fall on the point O in the direction LO. At O a perpendicular OP is drawn, here then the angle of incidence is POL = i. Now let us assume that an actual experiment has proved the direction of the ray in water after refraction at O to be OR, then the angle of refraction is ROT = r. To find the relation between these angles, let us suppose any circle to be drawn from O as centre, and the circumference to cut the ray before and after refraction at L and R. Now if the lines LS and RT be drawn from the points L and R respectively, so as to be both at right angles to the perpendicular at the point of incidence O, we shall find that SL bears to RT a definite proportion which is always the same for the same two substances. In this case the proportion will be as 4 is to 3. Now the lines SL and RT are called the *sine-lines* of the angles POL and ROT, or i and r respectively. Hence we may express this law of refraction more simply by saying that *for two given media the ratio of the sine of the angle of incidence to the sine of the angle of refraction is a constant quantity*. In this case, for air and water, the ratio as shown in the figure is 4 : 3. This ratio $\frac{4}{3}$ is called the *index of refraction* for air and water. For air and glass it is $\frac{3}{2}$. It must, of course, be fully understood that if the incident ray OL is in any other direction, OR will also be in a different direction, but if the corresponding sine-lines be drawn, their ratio will always be 4 : 3 in the case of air and water.

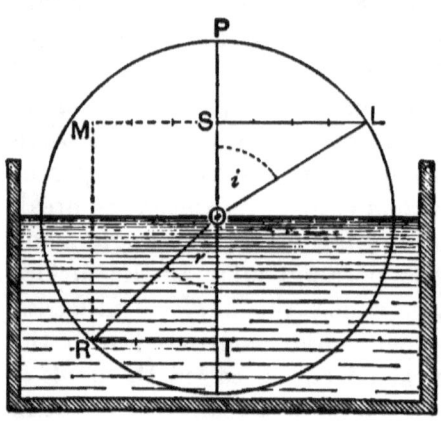

Fig. 50.

It is easily seen that if we know the index of refraction for two substances, we may easily find the direction of a given incident ray after refraction. Thus let LO be the given ray falling upon the surface of water at O. Describe a circle about O as centre, and draw OP perpendicular to the surface; draw LS perpendicular to OP, divide it into four equal parts; produce LS; measure off SM equal to three of these parts. Draw MR perpendicular to the surface, join R, the point where MR cuts the circumference, and O; then OR is the direction of the refracted ray, because $\dfrac{LS}{RT}=\dfrac{\sin i}{\sin r}=\dfrac{4}{3}$.

QUESTIONS ON CHAPTER XV

1. A square tank full of water is standing in the open air against a low wall. Explain why the shadow of the wall formed on the side on which the tank stands is narrower at the bottom of the tank than elsewhere along the wall. Illustrate your answer by a diagram.

2. A thick pane of plate glass is placed upon a sheet of white paper on the table, so as to leave a part of the paper uncovered. One of the two portions of the paper appears to be displaced and raised above the other. Will this part be that covered by the glass or the uncovered portion? Explain what you expect to see and give a diagram.

3. Explain clearly, with the help of a neat diagram, what is meant by the terms *angle of incidence, angle of refraction.*

4. What is the *cause* of refraction? By what experiment can your explanation be supported?

5. A person looking over the edge of an empty lock can just see the opposite wall down to the point where it reaches the bottom, but none of the bottom of the lock. When water is admitted a great portion of the bottom becomes visible to him though he has not altered his position. Explain this, with the help of a good sketch. Which of the experiments in this chapter closely resembles in its circumstances that described in this question?

6. A piece of glass is cut so that if a section were made anywhere across its length it would form an isosceles triangle, the angles at the base being each 75 degrees. Suppose that a ray of light passes in at one of the equal sides and goes out again at the other. Show by a diagram what you expect to be the path of this ray. How does the figure illustrating the action

of the burning glass in Chapter XIII. help you to answer this question?

7. If the velocity of light in air and water were known to you, how would this enable you accurately to draw the path of a ray of light which passes obliquely from air into water. Which of the diagrams given in this chapter will help you to answer this question?

8. It is said that in consequence of the refraction of the sun's rays by the atmosphere the sun apparently rises earlier and sets later than it would do if there were no air all round us. Explain this with the help of a good figure.

9. A sportsman is shooting at a fish swimming in the water. In what direction must he aim in order to hit the fish? Explain with the help of a neat sketch.

10. In looking through a window pane, do we see objects in their real position? Draw an accurate sketch in order to assist you in your answer.

CHAPTER XVI

DISPERSION OF LIGHT

Experiment 1.—Allow light from a candle flame, AB in Fig. 51, to pass through a very small aperture in a sheet of pasteboard, and to fall upon a screen of white paper.

Observe that an inverted *image* or picture of the candle flame will appear upon the screen. In order to understand how this image is produced, let us assume that instead of the complete candle flame only one of its points, for example the point A at the top, is sending forth light. This light would be sent all round the point in all directions, but only the ray AH, which reaches the paste-

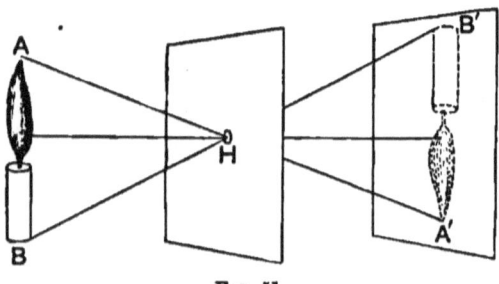

Fig. 51.

board at the spot where the hole is, could pass to the other side, and this would reach the point A′ on the screen; clearly at this point a small patch of light would be seen, which would be the "image" of the point A. The same reasoning would apply to the point B, and an image B' would be produced on the screen at the end of the straight line BHB'. In the same way an image of every single point composing the whole flame is produced at the end of the straight line passing from the point through H to the screen. Finally, the sum of all the luminous points, that is the whole of the flame, is reproduced as an "image" of the flame by the sum of all the images of the various points. We see also that points on the top of the flame must have their images below the images of points lower down in the flame, and that hence the image of the flame is "inverted."

Experiment 2.—Allow a beam of light to pass through a round hole in a sheet of stout pasteboard, and to fall upon a sheet of white paper or screen.

The waves of light which in this case produce the beam pass on in parallel straight lines, and hence produce on the screen a round spot of light (A in Fig. 54), which differs only in brightness from the remainder of the white surface, and is hence called a spot of *white* light. If the hole in the pasteboard were square, the spot of light on the screen would be seen to be square; it would invariably present in a more or less sharply defined manner the shape of the aperture, and is hence called an image of the aperture.

Experiment 3.—Hold a hollow triangular prism, as C in Fig. 54, in the path of the beam.

A triangular prism is a body which has three parallel edges, so that if a section is made perpendicularly to any edge right through the body, the surface of the section will be bounded by a triangle. In this experiment the prism is a hollow vessel formed by three oblong sides, and the beam of light passes through two of these sides. On looking at the screen we shall find that the round patch of light is still in its former place, or only very slightly displaced. Each ray of light which passes through the hole in the pasteboard reaches the nearest glass plate in a direction inclined to the perpendicular at the point of incidence, hence it is refracted towards the perpendicular, but on emerging at the other side of the same plate it is refracted away from the perpendicular and passes on in a direction parallel to its original one, though not in the same straight line as before. The same happens at the

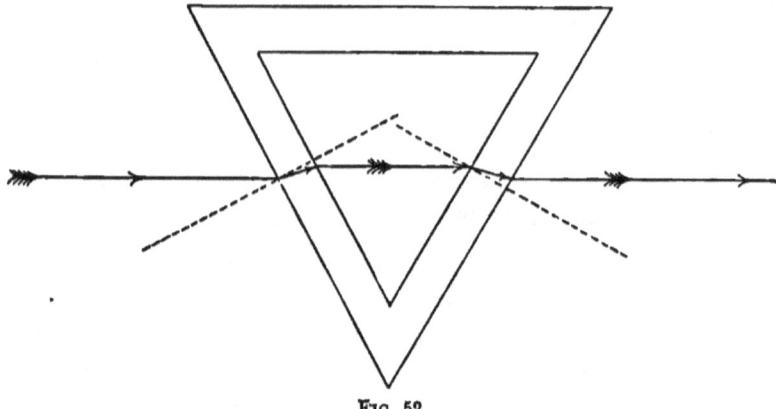

FIG. 52.

second plate, but as this is inclined in such a way that the ray on reaching it makes almost the same angle with the perpendicular, no very great displacement can take place, the less so as the inclination of the two sides to the ray is in contrary directions; all this will appear clear by studying Fig. 52, in which the direction of the ray of light before and after the repeated refractions which it has to undergo is shown by arrows. It follows that a *hollow* prism of glass or of any other transparent substance has no perceptible effect upon light which passes through it.

Experiment 4.—Repeat the previous experiment after filling the triangular space within with water and closing the prism.

CHAP. XVI   DISPERSION OF LIGHT   159

No white spot will now be seen on the screen, but the light which has passed through the *water* prism will be found to have suffered considerable changes. In the first place it is now turned away from its original path, and that which appeared before as an image of the round hole at the end of the straight path of the light through the aperture to the screen, is now somewhat higher up, and has lost its circular shape; it now appears as a short band with circular ends, somewhat of the figure which would be formed by placing a few shillings in a row, but each overlapping the other slightly. Secondly, the image which before was white now appears coloured, red being below and violet above; other colours may also be seen by close observation between the two, but a water prism is not very suitable for observing in a complete manner all that is to be seen in such an experiment. We have seen, however, in what way a triangular prism of any transparent substance acts upon a ray of light. It refracts the ray and indicates that white light is not simple, but is the result of a mixture of light of various colours.

The upward displacement, or "deviation" of the light from its former path will be easily understood from Fig.

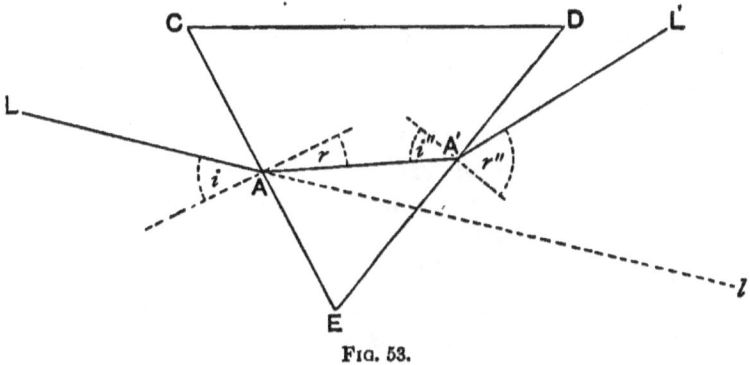

Fig. 53.

53, in which we suppose the prism to be glass throughout. At the point of incidence A of any ray LA, we suppose a perpendicular to be drawn to the side of the prism; the angle of incidence i is larger than the angle r, which the ray makes with the perpendicular within the prism, because glass is denser than air, and the refracted ray proceeds within the glass in a direction nearer to the perpendicular as well as nearer to the side CD, or "base" of the prism. Reaching the second surface at A' it is again refracted, but this time the angle of incidence i'' is smaller than the angle of refraction r'', because the ray leaves the glass for air, and since this is less dense the refraction is away from the perpendicular at A'. The ray thus makes a farther approach towards the base of the prism, and an image of the luminous point L is thus formed not at l, as would be the case if there were no prism in the path of the light, but higher up at L'.

Experiment 5.—Make a mark on the screen where the white image of the aperture is seen when no prism is used. Then cover the aperture first with a red and then with a blue piece of glass, and mark on each occasion the position on the screen of the circles of light, red and blue respectively, which are formed upon it.

On carefully conducting the experiment we shall see that the red light appears a little above the mark where the white light appeared when the prism was absent; this is a consequence of refraction solely, for if white light were not a mixture of various colours of light, still the white image would after refraction appear higher than the white image produced on the screen by direct unrefracted light as seen from Fig. 53. But using the blue glass we observe that the blue image appears upon the screen in a position decidedly above the red one. In other words, blue light is thrown more out of the straight path, or is more refracted, than red light. Hence the fact that light is a mixture is rendered manifest by the action upon white light of transparent substances such as water and many other media which exert different refractive actions upon the individual sorts of light in the mixture, and thus separating one colour from the other render each separately visible.

We have already seen that glass refracts more than water; it will hence appear probable that just as substances differ in their refractive power, so there may also be substances which separate the various colours better than water does. This kind of separation of white light into its coloured constituents is called *dispersion* (Latin, *dispergere*, to scatter apart).

Experiment 6.—Repeat Experiment 3, using a solid triangular prism of glass.

Observe that the result of this experiment is in a twofold respect different from those of the two preceding experiments; not only are the red and blue images of the aperture higher up, that is, farther away from the

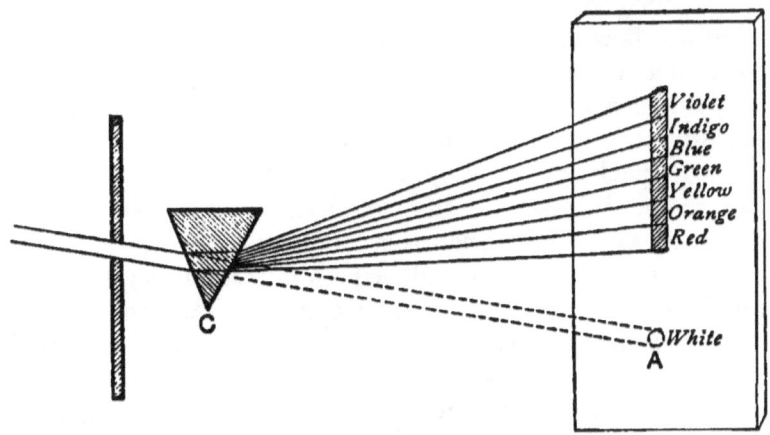

Fig. 54.

unrefracted white image, in other words, not only is the refractive action of glass seen to be greater than that of water, but there is also a much greater dispersion produced. A band of light, as shown in Fig. 54 makes its appearance upon the screen showing *red* at the lowest end, then *orange, yellow, green, blue, indigo,* and last at the extreme end *violet.* Whatever substance is used for producing dispersion of light, the colours mentioned are always produced in the same order; thus red is proved to be less refracted by any transparent substance than orange, orange more than red and less than green, and so on, violet being most refracted of all, or as it is called being more "refrangible" than light of any other colour. The band which appears upon the screen is called a *spectrum.* If the light, as in this case, is ordinary daylight, it is called the *solar spectrum.* The rainbow is a solar spectrum on a grand scale. It is the result of the dispersion of sunlight by raindrops.

Experiment 7.—Hold a burning-glass between the glass prism and the screen.

Observe that if the screen is placed at the same distance from the lens as is required for the experiments in which paper and tinfoil are exposed to the radiant energy of the sun, in other words, if the variously coloured light is brought to a focus on the screen by means of the burning glass, a speck of white light will make its appearance on the screen instead of the coloured band. The burning glass has as we have seen the property of collecting by its refractive action all the light which falls upon it; hence we see from the experiment that we may refract the various colours back again, so as to blend them, and that the result of the mixture will be white light.

If the light which reaches us from the sun, or any other luminous body, were caused by waves of a definite and uniform kind, water or glass or any other transparent substance could only produce upon the rays which propagate these waves one definite kind of action, namely refraction, and each transparent substance would then simply differ from others in the degree of this action. Now we have already seen that there are notes issued by various sounding bodies which differ in pitch, and that this difference of pitch depends solely on the difference in the number of air-waves which strike the ear in a second. Hence we conclude that colour must depend on the number of aether-waves which strike the eye in a second; and that white light is produced by a mixture of all kinds of waves, the red light being produced by longer waves than the orange, the orange by longer waves than the green, and finally the violet by the shortest waves of all. Substances like water, glass, etc., have therefore the effect upon light of retarding the longer waves less than the short ones; the shortest waves, those of the violet, being most retarded, are thrown out of their straight path much more than the longer waves, those of the red for example. Hence dispersion is due to the fact that white light consists of an infinite variety of waves sent forth in every ray and to the different refractive actions of every transparent substance upon waves of different kinds. It follows that, strictly speaking, the spectrum of white light is composed of an infinitely great variety of colours, each one being very little different from the next in order of succession; but a sharp separation is clearly impossible on account of the gradual transition of one colour into the next. The seven colours mentioned previously thus represent the principal groups which have been distinguished.

Experiment 8.—Fill the hollow glass prism with Carbon Disulphide and repeat Experiment 4.

Observe the great length of the spectrum produced as compared with that in the case of water or glass. If the spectrum is received upon a screen some 2 yards from the prism, the dispersion would lengthen the spot to a band nearly five times the length of the aperture. But it must not be forgotten, both in this and previous experiments, that the colours overlap considerably about the middle of the band and hence produce there a mixture. This is called an *impure* spectrum. In order to produce a pure spectrum more refined experimental arrangements and expensive apparatus are required; still, a careful attention to the spectra as seen in the experiments described will ensure results of considerable value.

Experiment 9.—On a black sheet of paper lay two small rectangular pieces of paper, B, Y, in Fig. 55, one blue the other yellow. Hold a slip of glass, as C in the figure, and bring your eye in a position to see one paper by reflected, the other by transmitted light.

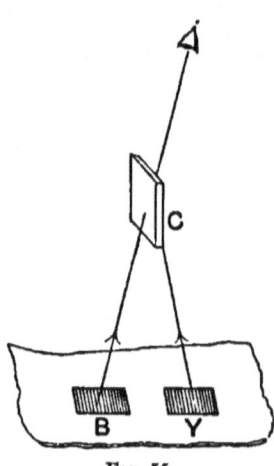

Fig. 55.

Here we see at the same time one colour direct and another by its reflected image, and they are arranged so as to overlap one another; the result will be that both colours will apparently disappear and in their place the result of their mixture will appear, and this will be white, or rather gray, which is white of a low luminosity. Thus not only a mixture of the prismatic colours will reproduce white, but white may also be produced by mixing together two colours only. Two colours which thus when mixed give white as the result of their mixture are said to be *complementary* to one another; here blue is complementary to yellow, and yellow to blue.

QUESTIONS ON CHAPTER XVI

1. Explain how a very small aperture in a sheet of paper may be used to produce upon a screen in a room an image of the landscape outside.

2. How far must the screen be placed from the aperture in Fig. 51 in order that the image of the flame may be three times the size of the flame itself? Why does the image become dimmer and less luminous the more the screen is removed from the aperture?

3. Why will a very small hole in a sheet of paper produce an image of an object, while a larger hole will produce an image of itself, when light is allowed to pass through it?

4. Explain how a hollow triangular vessel of glass differs, in its action on light, from a solid triangular prism.

5. Describe accurately what you see on a screen when a triangular vessel of glass, filled first with water and then with Carbon Disulphide, is placed in the path of a beam of light.

6. Describe and sketch what you expect to see when you hold a solid triangular prism of glass in a horizontal position close to the eye, and look through it at a small hole in a shutter through which sunlight is entering.

7. Describe fully how we may "analyse" light, and how we may prove that the various colours of the spectrum again give white light when put together.

8. Make a list of colours which you think will be complementary, and show how you would test your conclusions by experiment.

9. It is possible to recompose a coloured spectrum into white light again by means of two prisms, using one for decom-

posing the white light, and the other for recomposing the coloured light into white. Suggest how you would do it.

10. Suppose you wish to have your spectrum in a *horizontal* position on the screen instead of in the vertical one of Fig. 54; how would you have to place your prism? Show where, according to your arrangement, the red and the violet end of the spectrum would be respectively produced.

CHAPTER XVII

CHEMICAL ACTION OF RADIANT ENERGY

Experiment 1.—Dissolve a little common salt in water in a test-tube and a little Nitrate of Silver in another test-tube. Pour some of the salt solution slowly into the Silver Nitrate solution.

Both solutions are perfectly clear and transparent before we begin to mix them, but as soon as the smallest quantity of the salt solution has been dropped into the Silver Nitrate solution, a dense snow-white substance, looking like curd, makes its appearance in the silver nitrate solution, and in part sinks down to the bottom of the liquid, while a small portion will probably float about in the liquid.

It is clear that this white substance is insoluble in the liquid which surrounds it, or we should not see it as a solid substance; further, it must have immediately been formed by the action of the two solutions upon each other. Now common salt is a chemical compound of the metal sodium, with which we have already become acquainted in Part I., and a greenish-yellow gas, called chlorine. Both these substances are not only widely different in all their properties, but also differ so strikingly from common salt, which is nothing else but a compound of them, that this body is a beautiful illustration of the fact that when elements with different properties chemically combine the resulting compound differs as much from either of them as they differ from one another. Silver nitrate is a compound which results from dissolving pure silver in nitric acid.

Now if we were to separate the white curdy "precipitate" (Latin, *precipitare* = to throw down) from the liquid around it by filtration, which is, however, not our present object, we should find after evaporating the clear liquid that a well-known substance would be left in white crystals, called Chili Saltpetre, or Sodium Nitrate. Thus we may say that in mixing the two liquids chemical action has been started, and resulted in a mutual interchange of elements which may be represented thus—

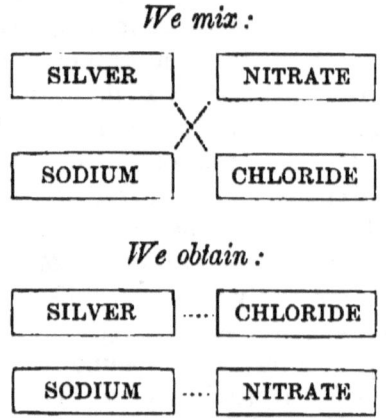

The white substance thus contains only Silver and Chlorine.

The action which takes place on this occasion is not only a type of thousands of similar actions with which the science of Chemistry abounds, but it also illustrates a very important law of chemical action, namely that when solutions of different compounds are mixed, chemical action will take place if any compound insoluble in the liquid can be formed by that kind of re-arrangement of the elements present which is here exemplified and is known as *double decomposition*; the compound thus formed will then be precipitated.

Experiment 2.—Make a strong solution of Hyposulphite of sodium and add some of it to the white precipitate.

On shaking the mixture a clear solution will be obtained. The silver chloride, though not soluble in water is soluble in water which contains Sodium Hyposulphite. We have every reason to believe that here again a new chemical action and not a mere solution has been brought about by mixing the solid silver chloride and the hyposulphite solution, for on now tasting a very small drop of the liquid we shall find it intensely sweet, while neither the silver chloride nor the hyposulphite have anything like a sweetish taste previous to their action upon one another. Further experiments would show that the sweet taste is due to a new compound, which can be obtained in crystals from the solution, and which consists of hyposulphite of sodium *and* silver, the silver having pushed out some sodium from the original hyposulphite, while the sodium thus displaced has seized upon the chlorine previously combined with the silver; after separating the sweet compound the liquid left consists of a solution of common salt.

Experiment 3.— Repeat Experiment 1, and immediately on obtaining the white precipitate close the test-tube by a stopper, and expose it to daylight in such a way that one-half of the test-tube is turned to the light, the other half away from it.

The portion of the precipitate which is exposed to the light gradually assumes a slaty colour, which becomes darker as the exposure continues and passes into a deep violet shade. That portion which has been turned away from the light will be found still to be white, and a more or less sharp boundary will be seen to separate it from the portion upon which the light has acted.

In this case no new substances have entered the test-tube during the experiment, and as light is not a substance, but consists of vibrations which have been started at the sun and propagated through the aether, the change which obviously has occurred can only be due to the fact that the vibrations of the aether have shaken the arrangement of elements which existed previously and have caused new arrangements amongst them, just as the fall of the waves on the seashore constantly rearranges the pebbles scattered on the beach. A dark coloured substance has clearly been formed, and we shall endeavour to lay hold of it separately.

Experiment 4.—Add some strong solution of hyposulphite of sodium to the darkened chloride of silver. Shake the mixture well and filter it.

A black powder will be left behind on the filter paper after the liquid has run through, and this, when the experiment is carried out on a larger scale than ours, may be collected, and is found to be pure silver in a very finely divided state. The action of light on the original precipitate of chloride of silver has produced a different compound of silver and chlorine, to which the name subchloride of silver has been given; it differs from ordinary chloride of silver in containing twice as much silver combined with the same quantity of chlorine. When the hyposulphite of sodium is now added, it of course dissolves as much of the original chloride as had been unacted upon by the light; and further, it breaks up the subchloride into ordinary chloride of silver, which it also dissolves, and silver, which is left on the filter paper.

Let us now suppose that some chloride of silver were spread in a fine layer over this paper and exposed to the light. In time the paper would become blackened, and if the paper were then dipped into hyposulphite of sodium no change would be effected, for, owing to the action of the light, we should now have insoluble silver spread over the surface. But if we were to cut out some figure from another piece of paper and cover the chloride of silver surface with it, the light would act only on the part cut out and blacken this part only. If now the surface acted on were dipped into the hyposulphite, this would dissolve away the silver chloride from every part except that upon which the light had acted — that is, where the figure had been cut out; hence a black impression of the figure would appear permanently printed on our paper.

On these experiments and the principles to which they lead, the important art of photography is founded. An elementary example of their application is given in the Appendix.

Experiment 5.—Soak a sheet of paper for a minute or two in a solution of common salt, dry it, and after dipping it into a nitrate of silver solution and drying it again, place upon different parts of the paper pieces of red, yellow, green, and blue glass. Expose the paper thus covered to sunlight for a few minutes.

Observe that the paper under the blue glass, in spite of its comparative darkness, will be as strongly blackened as those portions of the paper which were freely exposed, while the parts of the paper under the other coloured glasses show only very slight traces of the action of light. Ordinary light contains, as we have seen, blue light mixed with light of other colours; and as the action of both is the same, we are from our experiment entitled to conclude that it is the waves which give rise to blue light which are chemically most active. Indeed, if light is admitted into a space after the blue rays have been sifted out,—as, for example, by passing the light through an orange coloured glass,—scarcely any chemical action upon prepared paper will be observed. It is for this reason that the certain amount of light which photographers require in their "dark rooms" is generally admitted through orange coloured glass, which excludes the chemically active blue rays.

QUESTIONS ON CHAPTER XVII

1. Some common salt has been dissolved in a glass of water, but so little that it cannot be detected by the taste. How would you prove its presence in the water?

2. Explain fully what happens when a solution of silver nitrate is mixed with a solution of common salt. What elements take part in the changes produced?

3. What substances will, in your opinion, be respectively produced if a little dilute hydrochloric acid is poured upon a little caustic soda dissolved in water in a small dish, and a little nitric acid upon a globule of pure silver in a test-tube? (Try the two experiments.)

4. Describe fully the properties of hyposulphite of sodium as regards its action upon chloride of silver.

5. What difference does it make in the action of the hyposulphite if light has acted upon the chloride of silver? Describe fully the experiments which can be made in support of your answer.

6. How would you show by experiment that the different prismatic colours have not the same action upon paper saturated with chloride of silver? Which colour is most active? Which do you think is less active than any of the others? (Repeat the experiment again to test the latter part of the question.)

7. Suppose the screen in Fig. 54 were made of sensitive paper and a spectrum were projected upon it by a glass prism; describe and sketch the appearance of the screen after the light has acted upon it for a few minutes.

8. Give an outline of the successive steps which you would suggest for producing a permanent picture upon a glass plate of a star cut out of a piece of cardboard. (Read the "Hints" given in the Appendix.)

9. Mention from your own observation any change of colour of substances which you consider to have been caused by the action of sunlight.

CHAPTER XVIII

ELECTRICAL ENERGY

Experiment 1. — **Place a strip of zinc foil into a tumbler $\frac{2}{3}$ full of water, to which a small quantity of sulphuric acid has been added.**

Bubbles of gas will instantly be seen to collect on the surface of the zinc, to break away from it, and to rise to the surface of the liquid, while others again collect on the zinc to replace those which have risen. We already know from previous experiments (Part I., Chap. XVI., Experiment 5) that the result of this action between zinc and sulphuric acid is that hydrogen is given off; we know therefore that the gas now produced is hydrogen. We may, however, now direct our attention to the fact that the zinc visibly wastes away—it disappears as metal, and we may describe the action as a displacement of hydrogen, which is one of the elements of which sulphuric acid is made up, by zinc. The sulphuric acid is hence called in chemical language Hydrogen Sulphate; when its hydrogen has been removed in the manner seen in the experiment, the zinc may be said to be substituted for the hydrogen, and when the zinc has taken its place a white solid body is formed called Zinc Sulphate. This compound may easily be obtained by evaporating our liquid, when it will be found to be left in the dish.

Experiment 2.—Repeat Experiment 1, using a similar strip of copper foil and some dilute acid as before.

No action will be observed in this case even with the most careful attention; neither would any solid residue be left behind after evaporating the liquid, nor would the copper, if we were to weigh it before and after the experiment, be found to have lost anything in weight. We conclude from this that copper is not acted on by dilute sulphuric acid.

Experiment 3.—Place two strips, one of sheet zinc, the other of sheet copper, in the same dilute acid, when clear, and arrange them opposite and parallel to one another, so as to be everywhere 1 or 2 inches apart.

The action will be the same as if the copper were not present; hydrogen will be given off from the zinc plate as in Experiment 1.

Experiment 4.—Allow the two strips to touch one another, outside of the liquid, along the edges which project from it.

A striking change will immediately take place. Hydrogen will still be given off along the strip of zinc, but only in very small quantity, while a large quantity of hydrogen will be seen to bubble up from the copper strip. Nevertheless the copper undergoes no perceptible change, while the zinc is gradually wasted away as in Experiment 1; and, moreover, if the liquid were now evaporated, sulphate of zinc, not sulphate of copper, would be found to have been formed. As soon as the two plates are again separated and kept everywhere apart, as in Experiment 3, the bubbles of gas rise up only from the zinc strip, as before.

Experiment 5.—Remove the zinc from the liquid, and while it is still wet rub a little mercury over the whole surface, until it no longer appears gray but bright all over with the liquid metal. Now repeat the above experiments in the same order.

In the first instance it will at once be seen that scarcely any hydrogen is produced when the zinc strip is placed alone in the dilute acid. Zinc dissolves in mercury, forming a mixture of the two metals, or, as it is called, an "amalgam" of zinc; hence the strip is now said to be "amalgamated" zinc. Now, as mercury would not act by itself upon the acid, we cannot be surprised to see little effect produced by the zinc in its present state of solution, when clearly the zinc presents fewer points upon which the acid can act than before.

Next, the action will neither be increased nor diminished by the mere presence of the copper; but as soon as the copper plate touches the amalgamated zinc plate outside of the liquid bubbles of hydrogen quickly appear on the copper as before, but none appear on the zinc, although the zinc is still the metal which alone wastes away while the copper remains unchanged.

We shall be able to understand more about the effect of amalgamating the zinc later on, but we can already see that the unamalgamated zinc would go on wasting away during our experiments, whether the copper plate touched it or not, and during our experiments we should constantly have to remove it in order to prevent this waste; while the amalgamated plate only acts when we bring it in contact with the copper plate outside the liquid, and does not waste away during the time when no contact is made with the copper.

Experiment 6.—Attach by means of "binding screws" a few feet of copper wire to the projecting edge of each plate, and instead of placing the metals themselves in contact, connect them by pressing the free ends of the wires together.

The bubbles of hydrogen will again appear at the copper plate; and again, as soon as the wire ends are separated, all generation of gas will cease at once; nor will any action take place, although the ends of the wires be held together, if the wire be cut anywhere, or one of the binding screws be unfastened from either the zinc or the copper.

Experiment 7.—Wrap round the free ends of the wires, in succession, a little paper, silk, and cotton, and press them together; again, hold the ends apart by a piece of india-rubber or of wood.

In all these cases, as soon as any of the above-mentioned substances forms the link between one wire and the other, the evolution of gas will immediately cease.

Experiment 8.—Replace the copper strip by a second zinc strip.

The action will be the same, whether the plates are separated, or touch at the projecting edges, or are connected by wires, or the wires are connected by means of any of the above mentioned substances—hydrogen will be given off from both plates uninterruptedly until they are wasted away. If both plates are amalgamated, there will be scarcely any action, or if there is a small generation of gas, the action will be the same along both plates.

The experiments prove that there must be a connection *of a particular kind* between the two different metals, in order that the peculiar actions observed—the production of hydrogen on the copper plate instead of on the zinc plate, and its production on the copper even if none appears on the zinc—should take place. The connection—in this case the wire—is thus an important factor in the changes observed. It hence appears highly probable that some peculiar influence resides along the whole wire; or, in plain words, that something is going on in the wire while it is used in the manner described.

Experiment 9.—Wrap a long iron nail with paper, so that both its ends project a little; dip the nail into iron filings and remove it.

No iron filings will cling to it.

Experiment 10.—Wind one of the wires, either that attached to the zinc, or that from the copper, about twenty times in close turns round the part of the nail which is covered by paper. Press the free ends of the wires together, and dip the nail into the iron filings.

Considerable quantities of filings will cling to either end of the nail. If the ends of the wires be separated, the filings will drop and the nail will now be as unmagnetic as before. Thus the wire has, under these circumstances, while connecting two different metals standing in sulphuric acid, the property of making a piece of iron for the time a magnet.

Experiment 11.—Place an ordinary compass or any freely suspended magnet on the table,—for example, a magnetised knitting-needle (Part I. Chap. XXIV.). Hold the wire from the copper over and along the magnetic needle, and connect its end with that from the zinc.

The needle will instantly turn round, tending to place itself at right angles to the wire; indeed, after a few vibrations it will take up a permanent position more or less nearly at right angles to the wire. Thus the wire has the additional property of "deflecting" a magnetic needle from its ordinary position when at rest, which is in a line nearly north to south.

If the ends of the wires are separated, or some of the other substances mentioned in Experiment 7 are interposed, the needle instantly swings back to its former position, and points north and south again.

Experiment 12.—Repeat the preceding experiment with the wire coming from the zinc plate, taking care that if in the previous experiment the south pole of the magnet was nearer to the copper plate than its north pole, the south pole should now be nearer to the zinc plate than the north pole.

Observe that the deflection of the magnet is now in the opposite direction; if in the former experiment the north end was deflected to the west, it will now be deflected to the east.

No other conclusion can be arrived at from this whole series of experiments, but that when zinc and copper are placed in acid and connected by a wire, the wire exhibits very remarkable properties.

Some of these properties agree exactly with those which could be observed in a wire if electricity produced by friction (Part I. Chap. XXV.) were sent through it. Thus, if a needle is placed within a small glass tube and a wire coiled round the tube in the manner of Experiment 10, the needle will become magnetic and attract iron filings, if electricity produced by friction be sent through the wire. Hence *electricity* is said to be the cause of the new properties which we have produced in the wire by connecting it with the two metals. Again, if we suppose the two plates side by side in the tumbler suddenly connected by our touching each plate with one end of a wire, and we find that remarkable effects show themselves with equal suddenness in the vicinity of the wire, even where it is at some considerable distance from the plates themselves, we can hardly resist the belief that a *flow* of something is rapidly set up in the wire itself; because no change whatever is observed as regards the wire and its external properties, and yet it becomes suddenly endowed with a kind of energy which, as regards the motion which it gives to the magnetic needle at some distance from it, is very similar to radiant energy. It is in consequence of these properties and the way in which they are called into existence that we attribute them to the passage of an electric *current* through the wire. No evidence has as yet been obtained which would prove that anything like a flow or current really takes place in the wire; there are, however, a number of terms used in the science of electricity, connected with this idea of a current, which, having been found very convenient, are therefore universally adopted. Thus the wire itself, or any substance which connects the two different metals, is called the *conductor*. The whole arrangement through which electricity is supposed to flow is called the *circuit*; when the connection is interrupted anywhere, the circuit is said to be *broken*, when all is completely connected the circuit is *closed*. Similarly we speak of breaking (or opening) and making (or closing) the circuit. The two metals and the acid together are called a *voltaic cell* or *couple* (from the name of *Volta*, an Italian who first made experiments on current electricity); a series of such cells is called a *battery*. The two metals used in one cell are called the *poles*, the copper being the *positive* pole, the zinc forming the *negative* pole.

It will be proved by more delicate experiments in Part III. of this work that when the copper and zinc plates are immersed in the acidulated water, electrical separation takes place in the same way as when a glass rod is rubbed; the copper becomes charged with positive electricity, the zinc with negative electricity. The quantity of electricity in each metal is exceedingly small; it is many thousand times less, even when very large plates of the two metals are used, than the quantity of electricty in a rubbed glass rod. The small charge being diffused over a large surface, there is at no point of this surface a sufficiently large charge accumulated to traverse over the smallest air-space; hence no spark is produced. Yet even for the separation of such small quantities of electricity, and the production of the consequent visible motions of bodies,—in other words, for the work done by these electricities when separated,— force is required. To this force, which causes the separation of electricities, the name *Electromotive Force* has been given.

QUESTIONS ON CHAPTER XVIII

1. State in a few short sentences the observations made on the action of the two metals in the first five experiments.

2. Would pure water produce the same action as acidulated water? (Try the experiment.)

3. What do you expect to happen when the two metals are allowed to touch one another within the liquid? (Try the experiment.)

4. What advantage is gained by amalgamating the zinc?

5. May paper, india-rubber, or wood, be used as *conductors*? Why not? Describe experiments in proof of your answers. What substances appear to be conductors?

6. Would you expect moist silk or cotton to be available as conductors? (Try the experiment.)

7. Describe from memory the directions in which the magnetic needle has been deflected in the experiments made on the action of a current upon it.

8. Describe fully all the other effects of the current which have so far been proved by experiment.

9. As far as our present knowledge extends, what similarity and what dissimilarity appears to you to exist between current electricity and frictional electricity?

10. Explain fully the meaning of the following terms: *conductor, current, circuit, voltaic cell, battery.*

CHAPTER XIX

EFFECTS OF CURRENT ELECTRICITY

Experiment 1.—Set up a battery of six copper-zinc couples (Fig. 56) and repeat the Experiments 10 and 11 of the preceding chapter using dilute sulphuric acid as the liquid in the cells.

We shall at once see that the nail becomes much more strongly magnetic by using six cells than when we used only one; further, that the deflection of the magnet from its original position, in which it points north and south, will be considerably greater —that is, it will make a larger angle with the line in which it was at rest originally.

The arrangement used represents a battery of six cells or couples. The copper of the first cell is connected with the zinc of the second, the copper of the second with the zinc of the third, and so on. The pole of one cell being firmly attached to the opposite pole of the next, only the terminal strips of metal are free in such an arrangement,

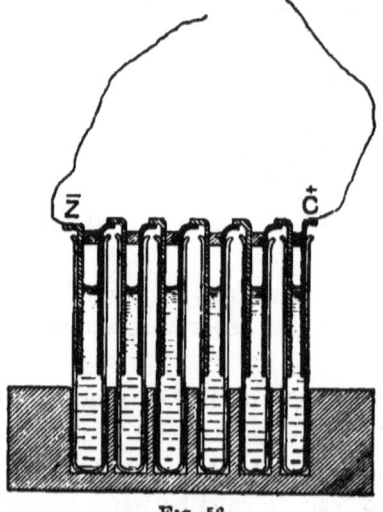

FIG. 56.

and they are called the "poles of the battery." When only two such cells are arranged in this manner, twice as great an effect is produced as by a single couple; the two plates of copper and zinc which are connected midway between the poles do not give electrical indications while they are thus connected—they serve as conductors of opposite electricities to the poles; hence it is found that the copper of the first cell is twice as strongly positive, and the zinc of the second twice as strongly negative, as when there is only one cell used. Similarly in our battery the copper of the first cell and the zinc of the last are six times as strongly electric as is the case in a single couple.

It has already been stated that when the poles are connected, either by direct contact or by two wires which are attached to both poles and brought together into direct contact, the circuit is said to be "closed." Now as the two kinds of electricity are called into existence at the same time, both of them must pass through the connecting body and recombine, and it is this movement which is called the "current," or "galvanic current" (from *Galvani*, the name of a celebrated discoverer in electrical science). It resembles the recombination of the two electricities produced by friction, when a positively charged body touches one which is negatively electrified, but it is very much weaker; on the other hand, it continues as long as the circuit is closed, because the electromotive force continues to bring about a constant separation of the two electricities as long as the poles are in conducting connection; the recombination of opposite frictional electricities, on the contrary, is properly a discharge-current, the two electricities which have combined being put out of existence. It follows that a galvanic current is, properly speaking, a double current, positive electricity flowing in one direction, negative flowing in the opposite direction; but for the sake of clearness and brevity the positive current is designated simply as the "current," if the direction of the current is spoken of, it being always understood that an opposite current, the negative one, simultaneously exists. Thus in one copper-zinc couple we should say that the current (that is, the positive electricity) flows from the copper pole through the connecting wire to the zinc pole, while negative electricity flows from the zinc through the connecting wire to the copper. For all practical purposes it is sufficient to say that, starting from the positive (copper) pole, the current passes through the wire to the zinc, and goes on through the zinc and liquid to the copper pole again.

Experiment 2.—Set up a magnetic needle so as to be easily movable, as in Experiment 11 of the last chapter, and note the direction in which the north pole is deflected when the current flows either from north to south or from south to north, (1) when the current is above the magnet, (2) when it is below the magnet.

We have already seen in the previous chapter that a freely supported magnetic needle is deflected by the current; the present series of experiments show us that the deflection obeys an invariable rule, and hence that from observing the deflection we are at once told in what direction the current flows. Thus we shall find that if the current flows above the magnet, from north to south, the north pole will be deflected to the east; if the current flows from south to north, above the magnet, the north pole is deflected to the west. Now let us assume that an observer were himself floating with the current—that is, head foremost with the stream as it were, and that being above it he would look down upon the magnet, then he would see in the latter case the north pole turning to west—that is, to his left; while in the former case, as he has to swim in the opposite direction in order to swim with the current, the north pole of the needle though deflected to the east would still be deflected to the observer's left, because his left is now on the opposite side. By considering the results of the above experiments for each case, we arrive at the following rule, which connects the deflection of the magnet with the direction of the current: Imagine yourself to be swimming in the current and facing the needle, then the north end of the needle will always be deflected to your left.

Since the needle would undergo no deflection if there were no current, it follows that a magnet capable of moving freely about a vertical axis—that is, a pointed support upon which it rests, or a thread by which it hangs, enables us to discover (1) the existence of a current near it, (2) its direction.

Experiment 3.—**Remove the acid from the battery used in the preceding experiment, and fill the cells with pure water instead. Try to repeat the experiments in which an iron nail was magnetised and a magnetic needle deflected.**

Observe that now scarcely any effect is produced in either case.

Experiment 4. — **Remove the water and use salt water instead of it. Repeat the two experiments.**

The magnetising and deflecting actions will now be very perceptible, yet they will be considerably less than in Experiment 1, where dilute sulphuric acid was used. It thus appears that the liquid used in the cell is of considerable influence upon the action of the current. The explanation of this is obvious. Very little chemical action is observable between zinc and pure water; a little more is seen to take place when the zinc is in salt water, for bubbles of gas rise in the liquid; much more still is noticed when the zinc is allowed to stand in dilute sulphuric acid. Hence we must conclude that electricity is developed by the chemical action between the liquid and the zinc. The electrical energy which, when conveyed along a wire, causes lively movements at more or less considerable distances from its source, cannot be created out of nothing. We have already seen that some substances—for example, a piece of phosphorus, sulphur, or charcoal, in combining with oxygen—that is, in consequence of obeying the simple impulse of one of the forces of nature, called chemical affinity,—give rise to considerable energy in the form of heat, which again may be converted, as is done in the steam-engine, into visible mechanical energy. As long as the phosphorus, sulphur, and coal have not yet entered into such combinations their energy is only potential, like that of a stone on the top of a precipice; but it becomes visible (or kinetic) energy in the act of combination with oxygen. Now, in the same way zinc enters into chemical combination with suitable substances, and the chemical energy which lay dormant until the zinc met the sulphuric acid, or other suitable body, at once becomes converted into the visible energy of heat, and electric energy when chemical action begins. Hence we may say that as coal is burned in the ordinary way when we want heat, so we burn zinc in the battery when we want electrical energy.

Hence we may also form some idea how it is that the hydrogen comes from the copper when the two metals are placed opposite to one another. The zinc clearly acts first on every one of those particles of liquid close to it, and, pushing their hydrogen out, takes its place. This hydrogen, instead of simply rising like ordinary hydrogen, may now be endowed with electrical polarity sufficient to push out the hydrogen not so endowed, which in the particles next to it is still in a state of combination with an oxide of sulphur to form sulphuric acid; the hydrogen in the liquid between the zinc and copper is thus pushed out gradually, by a series of decompositions and recombinations with the oxide of sulphur of the next layer, until finally, in the last layer next to the copper, the hydrogen, finding no longer any oxide of sulphur, is deposited as free hydrogen upon the surface of the copper, from which place it rises through the liquid.

Experiment 5.—**Fill the cells of the battery with dilute sulphuric acid and repeat the experiments of magnetisation and deflection; but instead of letting the wires touch, immerse their ends successively in water, salt water, and sulphuric acid, so that the liquid forms in each case a small layer between the wires.**

Observe that when salt water separates the poles scarcely any effect will be produced as regards magnetisation and deflection, and none whatever when water is used; even in the case of dilute sulphuric acid the action will be considerably diminished. Hence, just as we have seen that solids differ in their conductivity of electricity, so do liquids present striking differences in this respect. It follows that even where considerable chemical action appears to produce an appreciable amount of electrical energy, this energy may not be transmissible unless by solids or liquids which possess great conducting power, or, as it is usually also expressed, which present little "*resistance*" to the current. Thus, through paper, wood, india-rubber, gutta-percha, the skin of the human body, and similar substances, current electricity is so little able to travel that these bodies behave in reference to it like insulators. Pure water is a very bad conductor of galvanic electricity. Metals, carbon, and solutions of various salts and acids (sulphuric, nitric, and hydrochloric) concentrated or diluted are the only bodies available for conducting galvanic electricity. The solutions of salts and acids are very much better conductors than water; but even their conductivity is many hundred thousand and even many million times less than that of the metals. Of all known substances silver is the best conductor of galvanic electricity; copper stands next and very near to it in conducting power. Copper wire is preferably used for conducting galvanic electricity not only on account of its excellent conductivity, but also because the metal possesses great malleability and ductility—that is, the two properties of being easily hammered into any required shape, and as easily drawn out into wires of any required diameter.

QUESTIONS ON CHAPTER XIX

1. Describe fully the various effects produced by the small battery used in the experiments of the present chapter.

2. What is meant by the terms: poles of the battery; a closed circuit; an open circuit?

3. What is meant by the term *current*? Describe, with a suitable diagram, the direction in which the positive current is supposed to flow outside and inside a copper-zinc couple.

4. Describe fully, in a tabular form, the experiments made on the effect of the current upon a freely suspended magnetic needle, and state in each case the direction of the current, and the direction in which the needle was deflected.

5. Suppose the needle is above the current: what must be the direction of the current when the north end of the needle is deflected to the east?

6. Let the needle be below the current, and the north pole be deflected toward the east: what is the direction of the current?
    State the rule employed in the two last questions, and explain how you deduced your answers from it.

7. Explain fully the source of the electrical energy observed in our experiments.

8. What is meant by electrical conducting power? Mention good and bad conductors of current electricity. What substance is the best conductor of all?

9. How is it explained that the hydrogen appears at the copper plate of a battery when connection is made between its poles?

10. Suggest some way of comparing the strength of two currents produced by different batteries.

CHAPTER XX

FURTHER EFFECTS OF CURRENT ELECTRICITY

Experiment 1.—Set up a simple cell, but use a plate of iron instead of that of copper. Repeat with this couple the experiment on the deflection of a magnetic needle.

You will at once observe that the effect is a very faint one; the deflection of the needle will only be observed if it is delicately poised, and in that case will be considerably less than that produced by a zinc-copper couple. Now zinc is not the only metal ordinarily used for preparing hydrogen; iron may be substituted for it, and the generation of hydrogen is almost as vigorous as with zinc and dilute sulphuric acid. Hence the case is almost the same, as far as the production of a current is concerned, as when two plates of zinc are used. To prove this, let us assume that a plate of zinc is in a liquid upon which it acts chemically, and that an electrical separation, similar to that consequent upon rubbing a glass rod, is the consequence of this chemical action; now let a second plate of zinc be placed opposite to the first: the electrical difference set up by the first plate will clearly be set up to an equal amount by the second plate, and hence there can be no external current, because there is complete mutual neutralisation of the electrical differences. If iron acted exactly with the same chemical energy when displacing hydrogen from sulphuric acid as is manifested in the case of zinc, there would be no electrical difference—that is, no current would be produced, though there are two different metals opposed to one another, as in a zinc-copper cell; however, the chemical energy of the zinc in its action upon dilute sulphuric acid is somewhat greater than that of iron, hence the slight difference in electrical condition gives rise to a weak current, which produces only a small deflection.

Experiment 2.—Form couples of the following metals in succession, and observe the deflection produced in each case: zinc-tin; zinc-lead; zinc-copper; zinc-platinum; zinc-carbon.

The deflections will become larger and larger with each couple; in other words, a zinc-tin couple gives a larger deflection than a zinc-iron couple; a zinc-lead couple a somewhat larger deflection than a zinc-tin couple, and so on, until we arrive at the zinc-carbon couple, which gives the largest deflection of all.

These experiments prove that in producing a current it is important that one of the metals should be acted upon as little as possible or not at all. Hence it follows that the greater the disparity between the two solid elements with reference to their individual action upon the liquid, the greater is the difference in electrical condition produced, and hence the greater the current.

The experiments further show that even amongst those metals which have no preceptible effect upon dilute sulphuric acid a striking difference exists as regards their electrical behaviour with zinc. Zinc, which acts more powerfully on dilute sulphuric acid than the other metals of our series, is called the most electro-positive of these metals, while carbon, which has no action whatever on the same liquid, is called the most electro-negative substance in the series.

Hence we may arrange our series thus—

+ ZINC IRON TIN LEAD COPPER PLATINUM CARBON −

⟵⟶

where the direction of the current from one metal to the other within the liquid is indicated by the direction of the arrow.

The essential condition for the production of a galvanic current is thus the presence in a liquid of two different solids, one of which is more readily acted on chemically by the liquid than the other. Zinc and platinum, or zinc and carbon, are respectively the two substances best adapted to give a strong current. The former couple is employed in a battery called *Grove's battery*, the latter in *Bunsen's battery*, and also in the so-called *Bichromate battery*, the two latter differing, however, in the liquid employed. Several of these batteries are described and their mode of action fully explained in Part III.

The preceding experiments will also throw some light upon the difference between ordinary zinc and amalgamated zinc as regards their action in the voltaic cell. Zinc is prepared from an ore which contains many other substances associated with the metal, and some of these impurities remain in the ordinary metal as it is sold. These impurities are chiefly carbon, iron, etc., and when such a plate of zinc is immersed in dilute sulphuric acid, the particles of iron, etc., with the zinc, form numerous small voltaic circuits, and electricity will be generated and flow along the surface of the plate, at the expense of the zinc, thus diverting energy from the regular battery current and weakening it. But besides, this action, which has been called "local action," goes on even when the regular current is interrupted, and hence waste of zinc as well as of active liquid is occasioned, for from this action no useful energy is obtained. But if mercury is rubbed over the surface of the zinc (which must be done after the zinc has been dipped into acid to clean the surface) the mercury dissolves a portion of the zinc, forming with it a half liquid "amalgam," which covers up the impurities, and the amalgamated zinc then acts as perfectly pure zinc would act. Perfectly pure zinc might therefore of course be used instead of amalgamated zinc, were it easily obtainable, but this is not the case.

Experiment 3.—**Allow either of the poles of the small battery to touch the tongue first separately ; then place them side by side about half an inch distant from one another upon the point of the tongue.**

When separately allowed to touch the tongue, the metal ends will be found to be perfectly tasteless ; but as soon as they are both allowed to touch the tongue together, the latter serves to close the circuit, and this current, though very feeble, is sufficient to throw the nerves of taste which end in the surface of the tongue into a state of activity, and the sensation of a pungent taste is perceived. The sensation is, however, somewhat different at the two terminals ; at the negative pole the taste is acid, while at the positive terminal the taste somewhat resembles that of caustic soda.

In general, to pass the current through the human body, the parts of the skin which are brought into contact with the terminals of the battery must be moistened at least with water, or better with salt water to increase the conductivity. But even then the effect which is perceived is very small, and only a strong battery of a great number of cells will produce a "shock." It is in this general absence of sparks capable of passing through air, and the difficulty of obtaining "shocks" by current electricity, that its manifestations differ from those of frictional electricity. When a body is electrically excited by friction, the charge of each point of the same conductor being of the same kind, each charge repels those next to it; in other words, every small quantity of electricity, which forms a portion of the total quantity with which the conductor is charged, repels, and is repelled by, the remainder; each quantity tries to get as far as possible from the rest—that is, the whole charge will be found on the surface. This tendency of electricity to escape caused by electrical repulsion is called electric *tension*. Now in a galvanic current the quantity of electricity which is generated at any instant in one cell is many thousand times less than the quantity of electricity in a rubbed glass-rod, and this small quantity being diffused over a large surface, the tension of current electricity is exceedingly small, and the electricity quite incapable of traversing even the smallest air-gap, hence no spark is produced. Nevertheless a somewhat strong current is capable of producing very striking heating effects, and as the current flows continuously while an ordinary discharge of frictional electricity is instantaneous, the heating effects of current electricity are more easily observed than those produced by frictional electricity.

Experiment 4.—Set up a battery of two Bunsen's cells. Introduce between its poles a piece of fine Platinum wire.

The platinum wire will become red-hot, so that a piece of paper may be charred by bringing it in contact with the wire. If the platinum wire is so arranged as to be stretched over an ordinary gas burner, and the gas be turned on, it will be lighted by the heat of the wire. The experiments on the heating effects of currents may be considerably varied. We do not require a platinum wire to observe the production of heat; connection may instead be made between the two poles by means of a very thin copper wire 8 or 10 inches long. If a small strip of tinfoil, about 2 inches long and $\frac{1}{10}$ of an inch wide, be placed upon a wooden support, and first one terminal be pressed upon one end of the strip and then the other upon the opposite end, the strip becomes so hot that some part or other fuses. If the experiment be made in the dark it will be seen that this portion generally begins to glow with an intense light before it melts. A short bit of very fine iron wire is not only made white-hot, but is fused and begins to volatilise.

Experiment 5.—Set up a Bunsen's cell, and connect its poles by means of several feet of copper wire. Observe what deflection of the magnetic needle is produced by arranging above it one of the wires. Now introduce into the circuit a platinum wire as in the preceding experiment.

The deflection will become less while the wire becomes hot. The current is obviously weakened, for its action on the needle is now less than before. On the other hand some of the electrical energy due to the chemical action in the cell is clearly converted into heat. The weakening of the current is due to the *resistance* which the wire offers to the current, much as the friction between water and the interior of a pipe impedes to some extent the flow of the water through the pipe, or as a body pushed along a rough surface would not produce the same mechanical effect if it were to strike another body during its motion as that which it would produce if it were being pushed by the same force along a very smooth surface. In all these cases the heat produced is the equivalent of the energy expended in overcoming friction. So in current electricity, though it is with our present knowledge merely a convenient conception that a current is flowing through a wire, yet something does pass through the wire which is called electricity; and a longer wire, and, as the above experiments show, a thinner wire, must offer a greater resistance to the passage of that something than a shorter or thicker wire. Hence the transformation of a portion of the electrical energy into more and more heat-energy the greater the resistance opposed to the passage of the electricity is easily intelligible, as also the weakening of the current which accompanies the increase in the heat-production.

Experiment 6.—Set up a Bunsen's cell. Wind the wire from one pole several times round the smooth part of the iron of a file, and draw the end of a wire from the other pole over its rough surface.

The file forms part of the circuit, and as the wire passes over it the circuit is rapidly made and broken, and each break causes a spark at the point where the circuit is broken. The shower of sparks is due to red-hot particles of iron which are projected into the air. The appearance of the spark in this case is due, not to glowing air, as in the discharges of frictional electricity with high tension, but to glowing (or "incandescent") metal. A minute spark may easily be seen between the two terminals when contact is made and then again broken; the two terminals touch each other only at a metallic point which is sufficiently heated by the current to become luminous and to be volatilised. When the terminals are copper wires the sparks are small and appear bluish-green; iron wires produce larger and more brilliant sparks of a yellowish-red colour. When copper wire is used in connection with an iron file, as in the above experiment, brilliant yellow-red sparks interspersed with tiny bluish-green luminous points are observed, proving that the points of the teeth of the file and the end of the copper wire become incandescent at the same time.

When two pieces of some dense variety of carbon, which conducts electricity well, are attached to the ends of the connecting wires, and gentle contact is made between them, a small but very bright luminous point appears where the two pieces touch. When a strong battery of from 40 to 80 large Bunsen elements is used and two pointed pencils of carbon, connected with the terminals, are brought close together, light of most dazzling splendour is emitted by the two points, such as cannot be obtained from any other artificial source. This *electric light* is now frequently used for illuminating purposes on a large scale. At the present time, however, the electric energy required for its production is very rarely derived direct from chemical energy, but in the first instance from mechanical energy, which is generally supplied by steam-engines; but the mechanical energy of the steam-engine is after all originally derived from the potential energy of coal, which, in combining with oxygen—that is, in the act of burning—converts this potential chemical energy into the energy of heat, whence it is further transformed into the energy of visible motion.

QUESTIONS ON CHAPTER XX

1. Describe Experiments 1 and 2, and state the results derived from them. What practical application may be made of these results?

2. In a tin-lead couple, which is the positive and which the negative metal? In which direction would the current flow?

3. Arrange two couples so that the copper may be negative in one and positive in the other.

4. State and explain the various advantages derived from the amalgamation of the zinc used in a voltaic couple.

5. What is meant by electrical *tension*? Explain why galvanic electricity has so little tension compared with frictional electricity.

6. Describe fully the various effects of current-electricity demonstrated in the experiments of the present chapter.

7. Give an account of the various transformations of energy observable in these experiments.

8. How does a piece of zinc afford an instance of potential energy?

9. Point out the similarity between zinc and coal as regards potential energy. Could you produce electrical energy by means of coal? Explain the way in which you would proceed.

10. What is meant by *resistance* to a current? How is the resistance connected with the heat produced in a circuit? Will a thick platinum wire become hotter than a thin wire of the same metal if introduced in the circuit?

CHAPTER XXI

CHEMICAL EFFECTS OF ELECTRIC CURRENTS

Experiment 1.—Pour a little water into a clean evaporating dish and add a few drops of hydrochloric acid. Taste the liquid, and place a drop or two of it upon a piece of blue litmus paper.

The two operations will disclose two important properties of acids—first, that they have a peculiar taste, called a sour or acid taste, and secondly, that they change the blue colour of litmus paper to red.

Experiment 2.—Repeat the last experiment, using in succession sulphuric acid, nitric acid, and ordinary vinegar.

All these three substances have the same two characteristic properties in common with the hydrochloric acid. It does not follow, however, that if a substance does not taste acid, nor colours blue litmus paper red, that it is not an acid. In order to have a taste at all a substance must be soluble, for a solid body which is perfectly insoluble would produce no effect upon our nerves of taste, nor would such a body act on litmus paper. The colouring matter of litmus is a chemical compound extracted from several kinds of lichen which grow in Southern Europe, and the action which produces the change of colour consists in a chemical change of the blue compound brought about by the acid; it is a displacement of some elements by the acid and the formation of a new compound which is red. But such a change can in general only be brought about when the colouring matter is in solution, because only then can the mutual action which is caused by chemical affinity take place freely. Hence those solid acids which are insoluble will produce no change; with this reservation we may conclude, that if a body tastes acid and colours blue litmus red, it is either an acid itself, or some acid is present in the body tested, which may, however, contain other bodies as well.

Experiment 3.—Dissolve a little caustic soda in water. Taste the liquid, and place a few drops of it on the litmus papers reddened by acids in the preceding experiments.

The taste is not acid in this case; it is soapy and disagreeable, while its action on litmus paper is clearly the very opposite to that of acids, since it restores the original blue colour which this acid changed to red. Just as the above actions of acids are called their '*acid*' properties, so we speak of the actions of bodies which behave like soda as their *alkaline* properties.

Experiment 4.—Repeat the preceding experiment with caustic potash and ammonia.

These bodies are very similar in their action upon the organ of taste and upon red litmus paper. They have the additional property of producing when rubbed between the fingers a peculiar soapy feeling, which is due to a partial destruction of the outer layer of the skin; it is from this action that the attribute caustic (Greek *kauso*, I burn) is given to some bodies of this kind which are comprised under the general class of compounds called "bases," of which soda, potash, ammonia and a few others form the sub-class called "alkaline bases."

Experiment 5.—Add slowly and cautiously drop by drop from a graduated measure some dilute sulphuric acid to a small measured quantity of solution of soda, tinted blue with a solution of litmus, till the blue colour disappears and the liquid becomes red. Note the quantities of each liquid used.

If the experiment has been carefully conducted, a point will be reached when one drop of the alkaline liquid restores the blue colour, while another drop of the acid solution turns the blue liquid red again. It follows that there must exist a point at which the two fluids are so mixed as to completely *neutralise* each other—that is, at which the solution, if a drop of it were placed upon blue, and another upon red litmus paper, would effect neither the one nor the other. The acid and the base in the solution cannot when that point has been reached any longer exist as such in the liquid, and we must conclude that we then have before us not a mixture of the two bodies with their opposite properties, but a new chemical compound with properties different from either acid or base.

Experiment 6.—Mix some of the same dilute sulphuric acid as that used in the last experiment with some of the same solution of soda, making the proportion of each to be the same as that in which they neutralised each other, and evaporate the liquid to dryness in a dish.

A white solid body which is neither soda nor sulphuric acid will be left in the dish; it has no action on test-paper, nor will it taste either acid or soapy. The compound thus produced by the action of an acid upon an alkali constitutes what is called a *salt*. The salt before us is called Sulphate of Soda, or Sodium Sulphate.

Experiment 7.—Repeat the same operations of careful neutralisation and evaporation with a solution of soda and hydrochloric acid.

The white solid body left in the dish after evaporating the liquid will easily be recognised by its appearance and taste to be common salt, or, as it is called from its chemical constitution, Sodium Chloride.

Experiment 8.—Set up a battery of two Bunsen's cells and attach narrow strips of platinum to its poles. Dip the strips of platinum-foil into the two branches of a tube (Fig. 57) containing a solution of the sulphate of soda made in the preceding experiment, coloured blue with litmus solution.

It is clear that our arrangement forces the current to traverse the solution of sodium sulphate. As soon as the 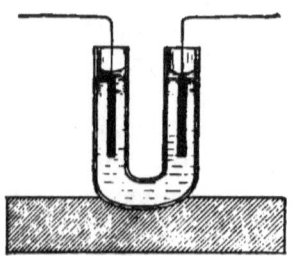 circuit is closed bubbles of gas are immediately disengaged from both platinum strips. The liquid round the strip connected with the positive (carbon) pole soon becomes red, while that connected with the negative (zinc) pole becomes of a deeper blue when litmus solution is used; when the solution has been coloured by a decoction of red cabbage, the liquid round the negative pole becomes green.

Fig. 57.

Evidently decomposition of the sodium sulphate has taken place; the salt has been separated by the current into the original acid and base—that is, into sulphuric acid and soda, the sulphuric acid appearing at the positive pole, the soda at the negative. Thus the current which owes its origin to, and is maintained by, chemical action in the battery is capable of doing chemical work outside the battery. Indeed, the most remarkable action of the current upon liquid conductors is the chemical decomposition which it produces in them. All liquids which are chemical compounds and are conductors of galvanic electricity are decomposed by it. This action of the current is called *electrolysis* (*electro*, and *lysis*, a loosening). The conductors (wires, plates, or the like) by which the current enters or leaves the substance to be decomposed are both called "electrodes" (*hodos*, way); the positive pole by which the current enters the substance is called "anode" (*ana*, up), and the negative pole where it leaves the substance the "cathode" (*kata*, down). The substance decomposed is called the "electrolyte."

Experiment 9.—**Remove the platinum strips from the liquid, close both ends of the little decomposing vessel with two fingers, and shake the liquid thoroughly.**

The original blue colour will be restored. This proves not only that the two separated substances, the base and the acid, have again entered into combination, forming the original salt, but they have been separated by the current in exactly the same proportion as that in which they existed in the compound.

In making Experiment 8 we saw bubbles of gas rising in the liquid at both electrodes. We may therefore conclude that another decomposition must be proceeding side by side with that of the salt, for the gas bubbles finally leave the liquid altogether and cannot form part of the sodium sulphate, which, as we have seen, is completely reformed when the liquid is shaken. As there is nothing else present except water and a little colour, it will be best to attempt the decomposition of water by the current, in order to obtain information on the nature of the gas or gases which appeared during the experiment with the sodium sulphate. Water is, however, a very bad conductor of galvanic electricity, and in order to increase its conductivity a small quantity of sulphuric acid is added to it, which remains unchanged while the water alone is acted upon by the current.

Experiment 10.—**Invert two test-tubes filled with**

acidulated water over platinum strips (Fig. 58) in a dish of water, and connect them with the poles of a battery of four Bunsen's cells.

As soon as the circuit is closed gas begins to collect in each tube, but the volumes of the gases collecting in the tubes are by no means equal; indeed, the bubbles which rise from the negative pole (cathode) are much more numerous than those disengaged at the positive pole (anode), and after a while, when some quantity of gas has been collected in each tube, it is seen that the gas from the cathode is about double that from the anode.

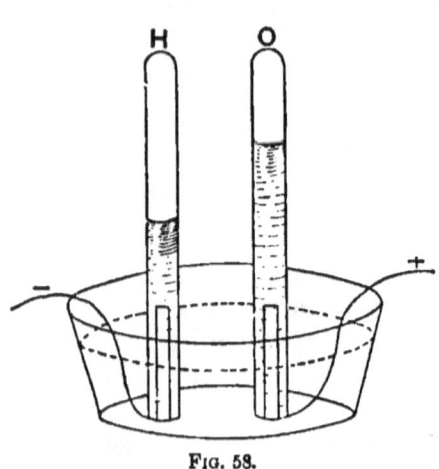

FIG. 58.

In order to investigate the nature of the gases collected, we stop the current when that tube which contains the larger quantity of gas is nearly full.

Experiment 11.—Cover the mouth of the tube with the thumb while still under the liquid, remove it, invert it while still closed, and bring a lighted taper close to and over the mouth. Now remove the thumb.

The gas rushes out into the flame which is above the mouth of the tube, takes fire with a slight explosion, and burns with a nearly colourless flame. From its behaviour, and our previous knowledge of hydrogen, we can have no doubt that the gas is hydrogen.

Experiment 12.—Remove the other tube in a similar manner. Prepare a glowing splinter of wood, and after withdrawing the thumb plunge the glowing wood into the gas.

The wood bursts into flame, but the gas does not catch fire. We know from our previous acquaintance with oxygen that this gas is oxygen.

Thus we prove by means of current electricity not only that water is a compound of the two gases, hydrogen and oxygen, but that two volumes of hydrogen and one volume of oxygen combine in the formation of water. Now careful experiments have proved that oxygen is bulk for bulk sixteen times as heavy as hydrogen; in other words, if our oxygen in the test-tube should happen to weigh 16 grains, an equal bulk of hydrogen would weigh 1 grain, and the hydrogen actually in the other test-tube would weigh 2 grains, because its bulk is twice that of our oxygen. It follows that 16 parts of oxygen by weight combine with 2 parts by weight of hydrogen to form water; or finally, that 18 parts of water contain 16 of oxygen and 2 of hydrogen in a state of chemical combination.

QUESTIONS ON CHAPTER XXI

1. Describe the principal properties characteristic of *acids* and *bases*. Enumerate as many members of each class as you know.

2. What is a *salt*? Mention as many bodies belonging to the class of salts as you know, and state in each case what you think to be the base and what the acid which together have formed the salt.

3. What is meant by *neutralisation*? Explain how common salt may be artificially formed.

4. Describe the effect of passing a current of electricity through a solution of sodium sulphate.

5. What proof have we that there is really a separation of the acid and the base taking place in the decomposition of sodium sulphate?

6. What would happen if after the liquid at the negative terminal became blue or green and that at the positive red, both the terminals were taken out and interchanged.

7. Explain the terms: *electrolysis, electrolyte, electrodes, cathode, anode*.

8. Describe fully the decomposition of water by electricity. How much hydrogen and oxygen, by weight and by volume, is held chemically combined in water?

9. How much by weight of hydrogen and oxygen respectively is there in 36 ounces of water?

10. While the decomposition of water is going on a magnet near the wire is less deflected than when no decomposition is proceeding. How would you account for this? (Test the fact by experiment.)

CHAPTER XXII

CHEMICAL EFFECTS OF ELECTRIC CURRENTS—(*Continued*)

Experiment 1.—Set up two Bunsen's cells and dip the platinum electrodes into the V-tube used previously, now filled with a solution of copper sulphate.

Copper sulphate is a compound formed by heating the metal copper with sulphuric acid; it is a salt in which the base is copper oxide and the acid sulphuric acid. The solution being blue, the effect of the current cannot be completely observed until after some time, when the circuit is opened by withdrawing the platinum strips from the liquid. We shall then find that the platinum strip connected with the negative pole is covered with a fine layer of copper. On the other hand, during the progress of the decomposition bubbles of gas are seen to rise at the positive terminal, which may easily be collected by slightly altering the arrangement, and could be proved to be oxygen. Sulphuric acid *and* oxygen are the products which appear at the positive pole, and it follows that in consequence of decomposition the solution is becoming weaker, for the copper would gradually be deposited on the negative platinum terminal until no copper in the form of copper sulphate would be left for further decomposition. It will be easily seen how important this fact is when we wish to discover how much copper a certain solution contains. By weighing our platinum strip beforehand, then withdrawing all the copper with the help of the current, washing and drying the strip, and finally weighing it again, we know at once the amount of copper which has been added, hence the amount present in the solution. As many other metallic salts behave exactly in the same way as the copper salt, we have thus obtained a convenient means of chemically analysing numerous substances, as far as regards the metallic constituent present in them.

Experiment 2.—**Interchange the electrodes, making the platinum strip with the copper upon it the positive terminal and the other strip the negative.**

The copper will now be deposited on the new cathode, while the sulphuric acid now disengaged on the new anode dissolves the previously deposited copper; this copper therefore gradually disappears again.

Experiment 3.—**Attach somewhat broad strips of platinum to a battery of one or two Bunsen's cells, and immerse them opposite to another in a weak solution of silver nitrate.**

Observe the beautiful growth of crystals which will shoot forth from the negative pole and spread towards the positive pole, bearing a strong resemblance to a vegetable growth; hence it is frequently called the "silver tree."

Experiment 4.—**Repeat the preceding experiment with solutions of tin chloride and lead acetate in succession.**

In a perfectly similar manner the metals tin and lead will respectively be deposited on the negative terminal and shoot forth as "trees" into the liquid; each metal has its own form of growth, and sometimes, particularly in the case of silver, the same metal exhibits different forms according to the strength of the solution and the strength of the current.

CHAP. XXII CHEMICAL EFFECTS OF ELECTRIC CURRENTS 209

Experiment 5.—Attach a small roll made of a few feet of fine copper-wire to the positive pole of a battery of two Bunsen's cells, place it in dilute sulphuric acid, and place opposite to it a platinum electrode attached to the negative pole.

We should expect a simple decomposition of water to take place, hydrogen making its appearance at the platinum strips and oxygen at the positive copper electrode formed by the roll of copper wire. Hydrogen will indeed make its appearance at the platinum strip for some little time, but no oxygen will appear at the positive electrode. The oxygen is no doubt separated, but it goes to form the base copper oxide, with which the sulphuric acid at once combines to form copper sulphate, which course of events is indicated by the fact that the liquid begins to be coloured blue, faintly at first, but more and more intensely as the formation of copper sulphate proceeds. Now the liquid is no longer a solution of sulphuric acid, but a solution of copper sulphate, and consequently the evolution of hydrogen at the negative pole stops, and metallic copper begins to be deposited on the platinum strip, the hydrogen now replacing the copper which is deposited. On the other hand, the sulphuric acid continues to combine with the copper oxide from the stock of copper kept at the positive pole to form copper sulphate, thus supplying the solution constantly with as much copper as is deposited on the negative platinum electrode.

Copper will under ordinary circumstances not dissolve in dilute sulphuric acid to form copper sulphate; thus our experiment proves that current electricity, so powerful in chemical decomposition, will under certain circumstances bring about chemical combination.

But the present experiment has another very important practical bearing. We may clearly substitute for the platinum strip any other conductor of the current—that is, any other metal—silver, brass, iron, also carbon, etc., and we shall obtain a deposit of copper which will completely cover the metal immersed with a layer of copper; or we may use a solution of silver nitrate or a soluble compound of gold, and thus deposit on any conducting substance attached to the negative pole a layer of silver or of gold, which adheres so firmly when the deposition of the metal has been continued for a short time that it can only be removed by hard wear and tear. In other words, the art of silvering or plating and gilding is founded upon the result of Experiment 5.

Experiment 6.—Attach successively a small bright silver coin, a piece of clean brass, and a piece of polished steel, by means of a fine wire, to the negative pole instead of the platinum strip used in the last experiment.

These substances will all in turn become covered with a bright layer of copper, which becomes thicker the longer the action is allowed to proceed, provided a sufficient supply of copper is attached to the positive pole. If the copper attached to the positive pole is weighed before and after the experiment, it will be found that it has lost exactly as much copper as would be found to have been deposited on the object attached to the negative pole.

When the layer of copper which is precipitated upon the object attached to the negative pole has been allowed to become sufficiently thick, it may be removed with ease, and will then form a more beautiful and faithful impression of the original than can be obtained by any other means. Those portions of the surface of the cathode (or object at the negative pole) which are raised of course appear hollow in the "cast," and *vice versâ*; but if the first cast be itself used as cathode, a second cast is obtained which is in every respect a most faithful copy of the original.

These facts form the foundation of the practical arts

called *electro-metallurgy* or *galvano-plastics*. It is thus that copper plates and woodcuts are multiplied. An engraved copper-plate or block of wood can only be used for a certain number of impressions, because the soft surface of the copper or wood is soon worn away; but by employing the galvanic current in the manner above indicated, a great number of casts may at once be obtained from the original plate or block, and these may be used like the original for producing good impressions on paper to any extent.

Experiment 7.—Set up the small battery of six zinc-copper couples, attach platinum strips as electrodes, and place them in acidulated water.

Bubbles of gas will appear at both electrodes, hydrogen at the zinc pole (cathode), oxygen at the copper pole (anode), as in the usual decomposition of water. But the action has no sooner commenced than it diminishes, and very soon ceases altogether. If the zinc and copper-plates are lifted out, wiped, and put into the acid again, the action begins anew, to diminish and altogether cease again.

It is not difficult to discover the reason of this. The current emerges from the battery at the anode, it enters at the anode the liquid to be decomposed, retaining the oxygen, and leading as it were the hydrogen to the cathode connected with the zinc pole, where it rises to the surface like the oxygen at the anode. Now the various chemical actions which we have observed both in the voltaic cells themselves as well as in the liquids in which the electrodes start chemical decomposition, must obey the same law throughout. The current which enters the liquid outside the cells at the copper pole clearly enters the cell itself by the zinc plate; hence the zinc plate is the anode *in* the cell and the copper the cathode, hence *in* the cell the oxygen is formed at the zinc, and the hydrogen is as it were conducted by the energy of the current to the copper-plate. This we have actually seen to take place in our first experiments made on current energy. Now the oxygen does not rise in bubbles at the zinc plate; it combines with it to form zinc oxide, which finds the sulphuric acid ready to form the zinc sulphate, which, as we have also proved, is gradually formed in the cell. If our little battery is filled with salt-water instead of acid, we should very soon find the zinc plate encrusted with a layer of white zinc oxide in the form of scales and flakes, there being no acid to combine with it. Now the hydrogen really escapes in bubbles at the copper-plate, but not all of it. Solid bodies have the property of condensing gases on their surface, which cling to them by the action of adhesion; they form an invisible fine layer, and our copper-plate soon no longer presents a surface of copper to the liquid; the points of contact of copper and liquid become less and less, and hence the quantity of electricity produced gradually becomes less also. Moreover, the film of hydrogen thus formed acts precisely like a metal with reference to the liquid which is interposed between it and the zinc, an independent current is started from a hydrogen-zinc couple, which, however, is directly opposite in direction to that of the zinc-copper current. The result is what has been seen in the last experiment; the current acts energetically for a few moments, but is rapidly enfeebled in such batteries which contain only zinc-copper couples inserted in dilute sulphuric acid. The removal of the hydrogen is effected by various mechanical and chemical means in the so-called *constant batteries*, to which class the Bunsen cell belongs. In this cell the hydrogen generated in the outer liquid (the sulphuric acid) passes through the porous cell and acts upon the nitric acid within the porous cell, partly decomposing, partly diluting it. It is the former action which gives rise to the various vapours accompanying this kind of battery.

CHAP. XXII CHEMICAL EFFECTS OF ELECTRIC CURRENTS 213

QUESTIONS ON CHAPTER XXII

1. State the results of Experiments 1 and 2. What gas is set free in this decomposition?

2. Hydrochloric acid is a compound of hydrogen and chlorine. Suggest some experimental arrangement for separating the two gases and investigating the nature of chlorine. (Try the experiment.)

3. In what respect will the Experiments 3 and 4 differ from one another in their results as regards the products which appear at the positive pole? (Repeat the experiments and pay attention to these differences.)

4. Describe fully how you would proceed to coat a large surface, say a pewter pot, with a layer of silver.

5. Describe Experiment 5, and point out in what respects it differs from the preceding experiments of this chapter.

6. What will happen in Experiment 5 when the supply of copper at the positive pole is exhausted? What appearance will be presented at the positive electrode when this happens?

7. Give a brief outline of the practical applications founded upon the facts brought to light by our experiments.

8. Explain why in a simple zinc-copper couple the hydrogen is seen to come from the copper when the circuit is closed.

9. What is meant by a constant battery? Describe the parts of some one constant cell with which you are acquainted, and its mode of action.

10. Place a piece of copper in some solution of copper sulphate contained in a porous pot; place the pot into a vessel containing dilute sulphuric acid and an amalgamated plate of zinc. Attach wires to either of the metals and close the circuit. Why does no gas appear at either metal? Is this a constant cell or not? (Try the experiment.)

CHAPTER XXIII

CHEMICAL ACTION BETWEEN METALS AND ACIDS

Experiment 1.—Take three test-tubes half-full of dilute sulphuric, dilute hydrochloric, and dilute nitric acids respectively. Throw into each test-tube a few scraps of copper turnings.

No action whatever will take place between the copper and the dilute sulphuric or the dilute hydrochloric acid, while lively action will be observed and gas will be evolved in the test-tube which contains the copper and dilute nitric acid. The gas is different from any with which we have hitherto become acquainted; it is a compound of nitrogen and oxygen, which may be received over water and examined. The liquid, moreover, assumes a blue colour, and contains nitrate of copper. By evaporating the liquid very gently over a "water bath" the copper nitrate may be obtained as a solid body. Thus the experiment proves that by the action of an acid upon a metal we may form a salt, provided that an interaction between the metal and the particular acid really takes place when they are brought together. The body formed is named after the metal and the acid jointly; so in this case, for instance, the body formed is called copper nitrate.

We have already seen that when zinc acts upon dilute sulphuric acid a salt called zinc sulphate is formed, and that in that action the gas evolved is hydrogen. Now while all acids, including the nitric acid used in our experiment, contain hydrogen, the salt copper nitrate, which is formed in this case, contains no hydrogen, and the gas produced, as has been stated, is a compound of oxygen and nitrogen. What, then, has become of the hydrogen contained in the nitric acid? Chemical investigation shows that the hydrogen is really removed from the acid by the copper, but that the gas removes in its turn some oxygen from the nitric acid and combines with it to form water, which remains in the test-tube, while the nitrogen and oxygen of the nitric acid which are left produce the gas which we saw evolved during the formation of the copper nitrate. Moreover, this gas is perfectly colourless, as can easily be seen if it is prepared on a larger scale and is prevented from coming into contact with the surrounding air, for it readily combines with oxygen; by the combination of these two colourless gases a new gas is formed which has a red colour, and makes its appearance even inside the test-tube, but more especially above it, in the form of a red cloud.

The behaviour of metals with acids, and the formation of salts, are of great practical importance, and should be carefully studied.

Experiment 2.—Repeat the preceding experiment, taking the same three acids and the metals in the following table in succession :—

You will observe the actions described in the table—

Metal.	Dilute Sulphuric Acid.	Dilute Hydrochloric Acid.	Dilute Nitric Acid.
TIN.	No action.	No action.	Red vapours evolved. A *white* oxide of tin formed.
LEAD.	No action.	No action.	Red vapours. A colourless solution of lead nitrate left.
ZINC.	Hydrogen evolved. A colourless solution of zinc sulphate left.	Hydrogen evolved. A colourless solution of zinc chloride left.	Red vapours. A colourless solution of zinc nitrate left.
IRON.	Hydrogen evolved. A bluish-green solution of iron sulphate left.	Hydrogen evolved. A green solution of iron chloride left.	Red vapours. A brownish-red solution of iron nitrate left.
MAGNESIUM.	Hydrogen evolved. Colourless solution of magnesium sulphate left.	Hydrogen evolved. Colourless solution of magnesium chloride left.	Red vapours. Colourless solution of magnesium nitrate left.
ALUMINIUM.	Scarcely any action (or very slow and minute).	Hydrogen evolved. Colourless solution of aluminium chloride left.	No action.
MERCURY.	No action.	No action.	Red vapours. Colourless solution of mercury nitrate left.
SILVER (Pure).	No action.	No action.	Red vapours. Colourless solution of silver nitrate left.
PLATINUM.	No action.	No action.	No action.
GOLD.	No action.	No action.	No action.

When the experiments are repeated on a somewhat larger scale, and the solutions are filtered and evaporated carefully, we obtain by the above actions the following compounds :—

By the action of dilute sulphuric acid : the *sulphates* of zinc, of iron, of magnesium, and (slowly) of aluminium.

By the action of dilute hydrochloric acid : the *chlorides* of zinc, of iron, of magnesium, and of aluminium.

By the action of nitric acid : the *nitrates* of lead, of zinc, of iron, of copper, of magnesium, of mercury, and of silver.

Tin is converted by the action of nitric acid into an oxide of tin, which itself partakes of the nature of an acid in its chemical character.

Experiment 3.—Put a few scraps of tin-foil into dilute hydrochloric acid and heat the liquid.

Hydrogen will be given off in a moderately large quantity, and a colourless solution of tin chloride is left; the salt itself may be obtained by gentle evaporation.

Experiment 4.—Put a few scraps of copper turnings into a test-tube, cover them with concentrated sulphuric acid, and heat gently.

A gas is given off which smells strongly of burning sulphur. The gas is indeed identical with the gas formed when sulphur burns, and is an oxide of sulphur. When the contents of the test-tube, after the action has been allowed to go on for some time, are poured into a little water and filtered, a clear blue solution of copper sulphate is obtained, which may be evaporated and will yield blue crystals of the substance.

The two last experiments prove that metals which will not readily combine with certain acids will do so when heat is applied. The action between the copper and the sulphuric acid in the last experiment is very similar to the general action between metals and nitric acid : the metal displaces hydrogen from the acid, and this hydrogen decomposes some of the sulphuric acid, robbing it of some of its oxygen to form water, thus "reducing" it to an oxide of sulphur, which is a gas, and contains less oxygen than the oxide which originally existed in the acid.

Experiment 5.—Weigh accurately a bright plate of zinc, place it in a solution of copper sulphate contained in a beaker glass, and remove it from time to time for inspection.

The zinc rapidly becomes covered with a red film, which gradually becomes thicker and drops from the plate in the form of a powder, which in time becomes a bulky red sediment. While this is going on the liquid gradually loses its blue colour and in the end becomes clear and colourless.

Experiment 6.—Weigh the zinc plate after washing it with clean water, and then carefully drying it with blotting-paper. Collect the red powder, dry it, and hammer it together upon an anvil. Evaporate the solution nearly to dryness.

The zinc will be found to have lost considerably in weight. The red substance will form a little mass, which can be easily recognised as copper. The solution will yield a white residue, which should form crystals of the same shape as those which were previously obtained by dissolving zinc in dilute sulphuric acid. Indeed, the result of the last two experiments is essentially that zinc displaces the copper in a solution of copper sulphate, and that zinc sulphate is formed while the copper is being deposited on the zinc; but as the zinc itself enters into solution the layer of copper soon forms a mass without cohesion, unlike the layer of copper obtained by the decomposition of copper sulphate, which is brought about by current electricity.

Experiment 7.—**Place a bright plate of copper into a solution of zinc sulphate.**

No effect whatever is observed; there is no mutual action between these substances.

Experiment 8.—**Place a strip of copper into a solution of silver nitrate, and after some time remove it.**

The solution, which was colourless at starting, will gradually become blue, an indication that copper has entered into combination with the acid; on the other hand, if the plate be carefully lifted out of the solution it will be found to be covered with a gray film, which is a deposit of pure silver. The copper will displace silver from a solution of any of its salts. If the experiment is inverted, and silver placed into a solution of a copper salt, no action will take place.

Experiment 9.—**Pour a little solution of nitrate of silver into a small china dish which contains a drop of mercury.**

After a little time a beautiful radiating deposit of pure silver will be found to have formed around the mercury, which displaces the silver from the compound and forms nitrate of mercury.

We thus learn that certain metals displace other metals from solutions of their salts, and by varying the experiments with a number of metals we shall find that the metals may be arranged in a series like the following :—

+ POTASSIUM SODIUM CALCIUM MAGNESIUM ALUMINIUM ZINC IRON LEAD TIN COPPER MERCURY SILVER PLATINUM GOLD −

which very much resembles the electromotive series

previously given. In this series the metal on the extreme left, potassium, has the greatest affinity for the acid with which any of those metals following it may be combined; hence it displaces all these metals, while it cannot itself be displaced by them. Thus aluminium would displace zinc from a zinc salt, but zinc would have no action on the aluminium salt, and so on. It will be seen that mercury, silver, platinum, and gold are easily displaceable by numerous metals. We indicate this difference in chemical affinity by saying in this case also that potassium, the metal at the extreme left, is more positive than all metals on the right side which follow it, that zinc is more positive than copper, and more negative or less positive than sodium, and so on.

QUESTIONS ON CHAPTER XXIII

1. Describe how you would form the sulphate, chloride, and nitrate of copper. By what external characters may these bodies be distinguished from one another.

2. Explain the difference in the action of sulphuric, nitric, and hydrochloric acids upon any metal with which they form salts.

3. In what cases does the action of sulphuric acid resemble that of nitric acid ?

4. Iron is placed in a solution of copper sulphate. What will happen ? What will be the colour of the solution at the end of the experiment ? State the reasons for your answer.

5. Select any metal of the series given in this chapter and any of the three above-mentioned acids, with which the metal forms no salt by direct action; show how you might form the particular salt composed of that metal and acid by displacement. (Try the experiment.)

6. What is meant by chemical affinity ? Give examples of chemical actions where strong affinity seems to induce the action, and of other actions where the affinity is very weak ?

7. Is the affinity between hydrogen and oxygen strong or weak ? Justify your answer by pointing to definite experiments which prove its truth.

8. When copper is placed in a closed bottle containing dilute sulphuric acid, no action takes place, however long the whole is kept in that state ; but when the same acid and metal are placed in a shallow open dish the liquid gradually becomes blue. How would you explain this ? (Try the experiment.)

CHAPTER XXIV

DEGREES OF CHEMICAL ACTION.   AFFINITY.   CHEMICAL ENERGY

Experiment 1.—Place a few thin slices of the metal potassium into each of two small china dishes. Leave one dish exposed to the air, and heat the other over a gas burner.

The heated metal will melt and its surface will soon become covered with a white crust; if the heating is continued the metal will burn with a bright flame, which has a violet colour and produces a white smoke. The process is simply an oxidation of the metal, and the white substance formed is an oxide of potassium. In the other dish the same combination is proceeding, though more slowly. If the back of the hand be from time to time brought into contact with the lower surface of the dish, it will be found gradually to become hotter as the oxidation proceeds. In this case, however, the substance gradually becomes moist, because a further combination is proceeding at the same time—namely, between the potassium oxide and the vapour of water present in the air, the result in the end being an oily liquid, called caustic potash, a compound which contains potassium, oxygen, and hydrogen; while potassium oxide contains no hydrogen.

Experiment 2.—Place a little potassium into a

"deflagrating spoon," heat it till it burns, and lower it quickly into a jar filled with oxygen.

Observe the brilliant flame due to the temperature of the potassium during its oxidation in pure oxygen being much higher than if air were employed, for air contains a large proportion of nitrogen, which not only takes no part in the chemical action, but absorbs heat, and hence lowers the temperature of those elements which do take part in the combination.

Experiment 3.—Take two rather large dishes which are half full of water, throw a small piece of the metal potassium into one, and a small piece of sodium into the other.

We have already studied the action of sodium upon water (Part I. Chap. XVI.), and have learnt that sodium decomposes water, liberating some of its hydrogen. It forms caustic soda by uniting with the hydrogen and oxygen left; in other words, given a suitable quantity of sodium and water, the result of the interaction between these two substances would be solely hydrogen and caustic soda. The same action takes place in the case of potassium; it decomposes water and forms hydrogen and caustic potash, the same body as is left in the dish exposed to the air in Experiment 1.

The hydrogen, however, which is produced by the potassium becomes so hot by the energetic chemical action that it burns with a bright violet-coloured flame. On the other hand the hydrogen produced by the action of sodium upon water does not burn, but the heat produced by the action is sufficient to melt the metal, which rolls about on the surface (the metal being lighter than water) in the form of a little molten silvery ball, which glows at one point or other if it happens to come in contact with the sides of the vessel, in which case friction increases its already high temperature.

We may conclude that if two similar chemical combinations are proceeding between different elements or compounds on one side, and the same element or compound on the other, and if in one case there is more heat produced than in the other, this extra production of heat, which is a form of energy, is due to a greater action of the force which is the cause of the combination. We have already learnt in Part I. that while the forces called adhesion and cohesion respectively enable different bodies, or parts of the same body, to hold together without altering their properties—that is, to remain what they are, a chemical change on the other hand invariably produces new bodies, different in their properties from the original bodies. It follows that chemical combination is the most intimate form of connection that we know, since in this case the actions between the elementary components must necessarily extend to the smallest possible particles of the body. It is on this account that a distinct name has been given to this uniting force, viz. *chemical affinity*—that is, a force which draws together the elements of the compound, firmly linking particle to particle, and which opposes itself to their separation.

But like most of the interactions between the matter of which the universe is built, and which come under our observation or are made the subject of experiment, the forces which are the original causes of these actions vary in degree, and these variations often depend upon very complex circumstances, which must be studied by experiment in each case separately, and taken account of in every single result or deduction to be derived from the application of scientific principles. Thus in Experiment 1 we saw the combination of potassium and oxygen obviously favoured—that is, the action of affinity strengthened, by the application of heat. Similarly, if mercury is heated to near its boiling-point —that is, to about 300° C.—it combines with the oxygen of the air and forms the red oxide of mercury; on the other hand we have in Part I. (Chap. X.) heated some red oxide of mercury above that temperature and have obtained mercury and oxygen—that is, the application of heat in this case overcame the effect of chemical affinity. Chemistry presents innumerable facts which prove that variations of temperature influence in different degrees the actions due to chemical affinity.

The principal conclusion to which the chemical experiments of the present and preceding chapters have brought us is one of the highest practical importance. The bodies which form the crust of our earth present an immense variety of different substances more or less exposed to the action of chemical affinity. A body like pure potassium, or sodium, or zinc, or a piece of coal, or a lump of fat, may be compared—in reference to chemical affinity—to a stone on the top of a perpendicular wall in reference to the action of gravity. The stone, from a comparatively slight cause, may be brought under the action of gravity so as to move; it will then possess visible energy of motion, which in its turn may be converted into other forms of energy, particularly into that form called heat. The piece of potassium, as long as it is removed from any element or compound with which it can combine, will represent potential energy; as soon as it is brought into favourable conditions for the action of chemical affinity, some of its potential energy is converted into visible energy, of which its rapid motion on the water may be taken as some indication, but the greater part is converted into the invisible motion towards each other of its own ultimate atoms and those of the body it combines with, and heat is the final form into which the energy of the motion accompanying the chemical interlacing of the atoms is transformed. Similarly zinc, in combining with dilute sulphuric acid in the voltaic cell, causes the visible motion of a magnet near it, and other forms of energy, such as heat and electricity, and does work in the disruption of compounds as in the decomposition of water. But neither potassium nor zinc are found in the crust of the earth as pure elements; like most of the substances of which this crust is made up, the affinities of the elements are already in most cases satisfied, and the potential energy which they possessed in their uncombined state has been converted into other forms many millions of years ago, when the rocks, which chiefly consist of salts, and complex compounds, and mixtures of salts, were formed from the elements which no doubt existed in a free uncombined state side by side, when the temperature of our earth was infinitely higher than at present. To obtain potassium and zinc as elements heat is required, and this heat is derived from the store of energy which we possess in our coalfields and forests. The potential energy of fuel is of the most essential importance to our existence, and it is clearly on the chemical affinity between the store of oxygen which we possess in our atmosphere and our store of fuel, chiefly in the coalfields of the earth, that our whole existence depends. By fuel we understand substances which are capable of combining with oxygen, and supplying us, as they so combine, with a large amount of heat of high temperature, which is employed either to warm ourselves and our habitations, or as an agent for generating mechanical effects in our various heat-engines. But just as energy had to be expended to obtain potassium and zinc, so energy had to be expended to obtain coal, which was originally combined with that very oxygen from the atmosphere with which we endeavour to recombine it when lighting a fire. The original energy which effected the chemical separation of the carbonic acid taken up by the plants came from the sun. The sun's rays, acting upon the leaves of plants, produce those decompositions which form fuel. The energy of the sun's rays has in fact been transformed into the potential energy of chemical separation. The rays of the sun, acting upon the leaves of plants in those remote ages when the plants which formed the coalbeds were in a state of growth, have laid up for man a stock of energy of inestimable value.

Food has the same origin as fuel, and its action partly rests on the potential energy stored in it, which enables it to combine with oxygen, and thus to produce the animal heat required. Plants serve to transmute the energy of the sun's rays into fuel *and* food. Animals consume this food and transmute it partly into useful work and partly into heat; but being heat of low temperature much of it is simply diffused, and thus rendered unavailable for further conversions into other forms of energy. An animal is in fact an engine, and just as an engine must be fed with fuel so an animal must be fed with food.

Experiment 4.—Collect the caustic potash formed in Experiment 1, and dissolve it in a little water. Pour a few drops of the liquid into a solution of copper sulphate.

A blue precipitate will be formed while, if a sufficient quantity of caustic potash solution has been added, the liquid will become perfectly clear and colourless.

Experiment 5.—Filter the clear liquid off and dip a plate of zinc into it. Collect the precipitate left on the filter.

No action whatever will take place, nor will any copper be seen on the plate. We conclude that all the copper previously in solution is now contained in the precipitate; further, that no free sulphuric acid can exist in the liquid, which thus must have combined with the potash to form sulphate of potassium.

Experiment 6.—Boil the blue precipitate with water and filter.

The blue precipitate will by boiling become black. It consisted of black copper oxide chemically combined with water; but the chemical affinity between these two compounds appears to be comparatively so weak that a moderate heat alone is sufficient to separate the water from the copper oxide. The latter may now be dried, and placed in a bulb tube. Hydrogen may then be passed over it (as in Experiment 7, Part I., Chap. XVI.), when pure copper will be obtained, the hydrogen combining with the oxygen of the oxide to form water. Here heat favours decomposition on the one hand and combination on the other at the same time.

We have thus obtained another method of removing a metal from a solution of one of its salts, and this method rests on the fact that potassium, as has been seen in the previous chapter, has greater affinity for the acid than copper.

It is upon these comparative degrees of affinity, and of the solubility of the compounds formed by the mutual action of bodies, that the whole so important branch of the science of chemistry called *chemical analysis* is founded. By studying these affinities we are enabled not only to discover the elements present in compounds, and the nature of the compounds themselves, but to separate them from one another when simply mixed, and finally to determine with the help of the balance how much of each is present in a compound or mixture.

Experiment 7.—Make solutions of copper sulphate, iron sulphate, zinc sulphate, lead acetate, mercury nitrate, and silver nitrate, in different test-tubes. Pour half of each solution into a second test-tube, and add to each liquid contained in one test-tube caustic potash, to the other ammonia, drop by drop.

The following small table will give the results of these experiments:—

Name of Salt.	Action of Caustic Potash.	Action of Ammonia.
Copper Sulphate.	Blue precipitate. Insoluble in excess of potash.	Blue precipitate. Soluble in excess of ammonia.
Iron Sulphate.	White precipitate (becoming gradually green and then brown). Insoluble in excess of potash.	As with caustic potash.
Lead Acetate.	White precipitate. Soluble in excess of potash.	White precipitate. Insoluble in excess of ammonia.
Mercury Nitrate.	Black precipitate. Insoluble in excess of potash.	As with caustic potash.
Silver Nitrate.	White precipitate. Insoluble in excess of potash.	White precipitate. Soluble in excess of ammonia.
Zinc Sulphate.	White precipitate. Soluble in excess of potash.	As with caustic potash.

It will be seen that the effects produced by caustic potash and ammonia are not only well adapted to distinguish the above solutions well from each other, but they may also be used to separate some of the metals from one another. Thus suppose we had a mixture of zinc sulphate and iron sulphate. By adding a small quantity of caustic potash we should obtain a mixed precipitate, the white precipitate of the zinc coming down with the green or brown precipitate of the iron. Now by adding "excess" of caustic potash—that is, more than is required for the precipitation—the whole of the zinc precipitated will dissolve in this "excess," and by filtering we obtain a clear solution which contains all the zinc, and an insoluble residue on the filter paper which contains all the iron previously in solution.

QUESTIONS ON CHAPTER XXIV

1. Describe what happens when potassium and sodium are thrown upon water. Point out what is similar in the two actions, and in what respect they differ from one another.

2. How does Experiment 3 prove that potassium is more positive than sodium?

3. When magnesium wire is burnt in air a very bright flame is produced; when burning magnesium wire is lowered into a bottle in which a little sulphur is boiling the flame becomes still brighter. What conclusion would you draw from this fact? (Try the experiment.)

4. How is chemical affinity different from adhesion and cohesion? Give examples of the actions of these three forces.

5. Where is the potential energy of a tallow-candle derived from? Into what kinds of energy may it be transformed?

6. Describe three different modes of obtaining pure copper from a solution of copper sulphate.

7. What colours have the solutions of zinc sulphate, iron sulphate, copper chloride, iron chloride, copper nitrate, iron nitrate, and silver nitrate respectively?

8. Mention the metals which are precipitated by caustic potash and redissolved in excess of potash?

9. How would you separate a silver solution from one of iron?

10. How would you separate a lead solution from one of mercury?

APPENDIX

HINTS FOR PERFORMING THE EXPERIMENTS

Chapter I

Experiment 1.—Sewing cotton may be used for this and the following experiments; still better is the kind of silk thread called "tailors' twist," being both stronger and more flexible than cotton thread, which is useless for suspending larger masses, as in the experiments further on. The thread used must be thin, and soft enough to show the pull, and form a straight line, when the smaller weight is attached. If possible, cylindrical weights should be used, and not the ordinary flat-shaped ones.

Experiment 6.—The clay ball must not be wet, nor dry. A little Stourbridge clay, moistened with very little water, so as just to make it cohere, should be formed into a round ball of about an inch, or even less, in diameter. It must be kept for a little while till no trace of moisture is seen on the outside before using it. If too moist, it will stick to the table and spread out upon it when it falls; if too dry, it will either not be flattened or fly to pieces.

Chapter II

Experiment 1.—A straight stick, having marks or notches at distances of 1, 2, 3, etc., feet from one end, will do as well as a foot-rule.

Experiment 3, etc.—It is best to have a block about 1 foot square and 4 or 5 inches thick, roughly sawn at one of the square sides, and nicely planed, or better polished, on the other square side. The board should also have one side rough, the other well planed or polished. The board should be 1 foot wide at least; better still about $1\frac{1}{2}$ foot; it must be made of well-seasoned wood, and not used if warped.

The weight used should be as flat a one as can be had; a tall one will of course topple over, instead of sliding with the block, when the board is raised. If possible a square flat block of lead should be cast for the experiments and used. A baking-tin of suitable size can easily be purchased from dealers in tin-ware, and old lead, some 15 or 20 lbs., may be melted in it and then turned out when cold, or left in the dish for use. The tin may be secured upon the block by two or three tacks driven into the wood, so as to prevent the tin from sliding.

Chapter III

Experiments 1 and 2.—Stout drawing-pins with broad heads, and about $\frac{3}{4}$ inch long, are very suitable for these experiments. The wooden scale should be placed as near to the tacks as is possible, without interfering with the fall of the weights; it serves not only as a measure of the height, but also as a guide for dropping the mass with greater certainty upon the head, so as to drive it straight into the wood.

Experiment 3.—The piston, or working handle, of an ordinary glass syringe, such as is sold by chemists, will serve well for this experiment. If a glass test-tube is used, it must be specially made from stout tubing; the ordinary test-tubes are far too thin to be used with safety in this experiment. A tube of copper, closed at one end, is best, but does not allow what is going on in it to be seen so well as a tube of glass. A piece of paper should be folded into a thick, narrow band, which is put round the tube; by this the tube is held while it is heated. Special "test-tube holders" are, however, sold for this purpose by the dealers.

Chapter V

Experiment 1, etc.—The frame and lever, Fig. 11, should be made of very hard wood by a skilled joiner. The whole should be well planed, and the holes bored neatly, and clean through. It is possible to dispense with the frame altogether by suspending the lever with the help of a strong, but not too stout, piece of cord, which is passed through the middle hole and fastened to some suitable support.

The weights are suspended by short pieces of thread simply passed through the holes and tied to the weights. A neater way is to pass a short piece of stout, straight wire through the hole, which projects a little at each end, and suspend the weights by a piece of thread having small loops at each end, which are hung upon the projecting ends.

Chapter VI

Experiment 1, etc.—A strong wooden support is indispensable for the experiments with pulleys; it is also useful for many other experiments. It should be made in the form of a rectangular or nearly square frame, a broad solid board about 3 feet long and 9 or 10 inches wide forming the foot. The two uprights and the crossbar should be of hard wood and about an inch square at least. If the whole can be made somewhat larger, without causing inconvenience in other respects, it is so much the better. Into the upper crossbar a number of small iron or brass hooks, which may be bought at any ironmonger's, will have to be screwed. A hole smaller than the screw is first made with a bradawl, and the screw is slightly greased with tallow before it is screwed into the hole. It is desirable to arrange each system first on the table, placing the pulleys flat upon it, then arranging the cord, and after measuring the required distances between the different portions to bore the holes in the crossbar and fix the screws accordingly.

The best and most flexible cord for our present purpose also is the kind of stout black silk thread, called "tailors' twist," used for making button-holes.

The pulleys must be well made and of brass, to be of any use for finding the mechanical advantage of each system from the experiments, with some approach to accuracy. They are not very expensive, and may be readily obtained from a dockyard model dealer. The diameter of the sheave should not be less than 2 inches.

For the reception of the mass used as Power, it is best to employ a small scale-pan of brass suspended by three short silk threads knotted together, and having a brass hook for attaching it readily to a loop at the end of the cord. The weight of the scale-pan must of course be taken into account when reckoning up the mass employed for producing equilibrium.

The more complex systems should first be completely "laid out" on the table in the proper connection of the parts, except the Power and Weight to be used; the most suitable way of suspending the whole to the frame will then readily suggest itself. The suspension itself is then *gradually* proceeded with, taking care to fix any loose end of the cord temporarily to a hook in the frame, until it is wound round the pulley and is ready for being attached to any heavy mass; the cord is then loosened, held a few inches from its free end firmly with the left hand, while the scale-pan and the weights, which are kept ready near the frame, are placed in position.

If the whole is not carried out with care and thoughtfulness, much time will be wasted and numerous collapses will occur, to the detriment of the apparatus employed.

In consequence of friction the relation between the Power and Weight, as theoretically calculated, will within certain limits differ from that found by experiment, especially in the more complex systems. The difference may be reduced to a minimum in the following way: When the system is at rest pull the cord which carries the Power gently, first a few inches downwards, then a few inches upwards; note whether the motion in one direction requires more effort on your part than in the other; if so add to, or take away a small mass from, the Power until the system moves both ways with equal ease.

Chapter VII

Experiment 4.—The pulley used in this experiment must be clamped very firmly in the fork of the kind of stand called a retort-stand, which is used for many chemical experiments. It is still better to use a "stand-pulley," in which the pulley is attached to a solid brass rod capable of sliding with little friction in a brass tube, which is fixed upon a broad metal foot. Such a pulley permits a nice adjustment of the cord, carrying the mass in a position exactly parallel to the plane. A clamping screw holds the rod at any height to which it is raised.

Chapter VIII

Experiment 1, *et seq.*—The table should be so placed that a free space of 6 or 7 feet at least is obtained on its left-hand side. The blocks should be made of some hard wood, and be of a cubical shape, the smallest having an edge of 5 inches, the next of 6·3 inches, and the largest of 7·2 inches. If made of the same kind of wood, their masses will then be approximately in the ratio of 1 : 2 : 3.

The blow can be better regulated if a croquet-mallet of average size is used instead of an ordinary hammer.

Experiments 4 and 5.—A substantial small cannon, such as is sold in the better class of toy-shop, should be selected, having a bore suitable for ordinary shot of the largest size. The leaden target can easily be cast from a piece of lead, of about the same weight as the cannon without its carriage, by melting the metal in a ladle and pouring it into a small shallow tray, such as the cup or lid of one of the tin canisters used for various articles sold by grocers. Both target and gun are suspended by stout copper wire from the horizontal support described above (Chap. VI. Expt. 1). The copper wires must not be so long as to allow the two bodies to strike one another while swinging to and fro. This point must be attended to carefully; free motion of either body must be ensured before carrying out the experiment. Care is also required to make sure that the bullet really hits the leaden target. This difficulty might be got over by having an iron target made, which could be made larger for the same weight. An iron target has also other advantages, but is not so easily procured as a leaden one.

Care must be taken that the shot used fits exactly the calibre of the gun; if too small, no attempt should be made to wrap it into paper or linen to make it fit; but a mould must be procured and suitable bullets cast for the experiments, otherwise they will end in failure.

Chapter IX

Experiment 1.—The thermometer should have a cylindrical bulb, as thin as obtainable, and be kept ready so as to be immediately applied to the part of the wood immediately surrounding the whole.

Experiments 2 and 3.—If thin laths, almost as thin as veneers, can be obtained, they should be very firmly fastened together by a few nails at the corners, so as to form a solid block about an inch or more thick, and 4 or 5 inches square. The same target would do for both shots by using the space near one end for Experiment 2, and that near another end for Experiment 3.

The charge of powder must be very small for Experiment 2, so that the space in the bore occupied by it, and by $1\frac{1}{2}$, 2, $2\frac{1}{2}$ times this charge in Experiment 3, should bear only a small proportion to the length of the gun. If this is not carefully attended to a serious accident may occur by overloading.

Experiment 4.—The clay should not be too soft. With a certain consistency of it, the depth can be easily measured by means of a pencil.

Chapter X

Experiment 1, *et seq.*—The bullets used for these experiments are of two kinds, of lead and of plaster of Paris. Either kind is cast in a common bullet-mould, easily obtained from an ironmonger or a gunsmith. A fine copper wire, about 2 inches long, is bent at one end into a small hook, and the other end is twisted into a small spiral, which is pushed into the hole of the bullet-mould before pouring the lead in ; the spiral serves as screw which fastens the hook to the bullet. Plaster of Paris, which is sold as a dry white powder, is made into a thin paste with water, and is poured into the mould after inserting the hooked spiral into it as before. After the lapse of two or three hours the mould may be opened and a hard bullet will be found, suitable for the experiment.

The thread for suspending the pendulums should be of very fine silk; its length for the longer pendulums used about 20 inches. It is necessary to measure out the lengths required with great accuracy. The object of suspending the bullets by two threads of equal length, as in the figure, is to confine the vibrations to one plane, which is difficult if the pendulum has only one thread of suspension.

Experiment 3.—The paper disc should have its diameter about $\frac{1}{10}$ of an inch smaller than that of the coin; it should be placed carefully upon the latter, so that the centres of the coin and the paper should coincide. The coin should be held gently with the thumb and the forefinger at opposite points of the edge, and dropped without any jerk.

Experiment 5.—It will be well, for this experiment, to suspend the shorter pendulum within the longer, the two middle hooks in Fig. 30 carrying the short pendulum, and the two outer hooks the long one. It will then be easier to start them at the same angle with the vertical, and to observe their simultaneous arrival on the same side.

Experiment 6.—The wire should be ordinary piano wire, not too thick; it must be well stretched after being fastened to one of the nails, or still better screws, which are fastened to the board, and then twisted while being kept stretched, round the other screw. If the wire is too elastic, it will be advisable to heat those portions which are to be twisted round the screws in a Bunsen gas-flame, this will render them more pliable.

Chapter XI

Experiment 1.—The glass tube should not be less than 3 feet in length, and from $\frac{1}{4}$ to $\frac{1}{2}$ of an inch in diameter; the wall of it should not be too stout. It will be best to try several tubes of various dimensions as regards length, thickness, and size of bore. The ear should be applied pretty close to the tube, as the sound of the tube will in any case be rather faint on account of its small surface; indeed perfect silence is necessary for the success of the experiment.

The vibrations may be well seen by holding the tube on a level with the eyes and perfectly horizontal. For rendering the motion at different points still more perceptible, devices, which are called *riders*, are used; these are small pieces of paper about $\frac{1}{8}$ of an inch wide and $\frac{3}{8}$ of an inch long, bent at an acute angle and placed astride the tube. Such a rider remains almost unmoved at the points A or B in Fig. 32, that is, upon the nodes, if either B or A is held between the fingers, but is at once jerked off if placed in the middle of the tube.

The musical note produced may be heard much more distinctly by tying one end of a thread about 2 feet long near to one end of the tube, and allowing the tube to hang freely in a vertical position in front of the body; the other end of the thread is wound round the tip of the forefinger and pressed by the latter into the ear. If the tube is now struck with a finger of the hand which is free, a very loud musical note will be heard.

If two threads, equally long, are similarly tied to the tube at the points A and B ($\frac{1}{4}$ of the length from each end) and the other ends of both threads are wound round the same forefinger and pressed into the ear, the whole will form a kind of triangle, of which the tube forms the horizontal base. The tube is allowed to hang in front of the body, and after placing riders in the middle and near the ends, it is struck near to the point A or B, when the relative motion of the different points of the tube may be well seen and the note heard at the same time.

Experiment 2, *et seq.*—For this and the next experiment it is very desirable to use pretty large forks, and it is essential to have at least two, of which the vibrations are as 2 : 1, and besides a fork sounding middle C, that is, making 256 vibrations in a second.

Experiments 4 and 5.—An empty cigar box, whose lid has been removed, makes an excellent sounding-box if placed

APPENDIX

bottom upwards on the table, and will show the effects of resonance much better than the table itself.

Chapter XII

Experiment 1.—This tube may be made of tinplate by any tinsmith, at a small expense. The parts A and B should be each at least 2 yards long and 4 inches in diameter. The conical or rather funnel-shaped cap C should be about 1 foot long, and the narrow aperture $1\frac{1}{4}$ inch in diameter. The parts should be made slightly tapering, so that they may be put together like a stove-pipe.

To fill the tube partly at least with smoke, burn a little touch-paper within the wider end while holding the tube in a slanting position, the narrow end uppermost. Touch-paper is easily made by dissolving a teaspoonful of saltpetre in a tea-cupful of hot water, dipping some stout clean blotting-paper in the solution and then allowing it to dry. The paper produces much smoke in burning but no flame.

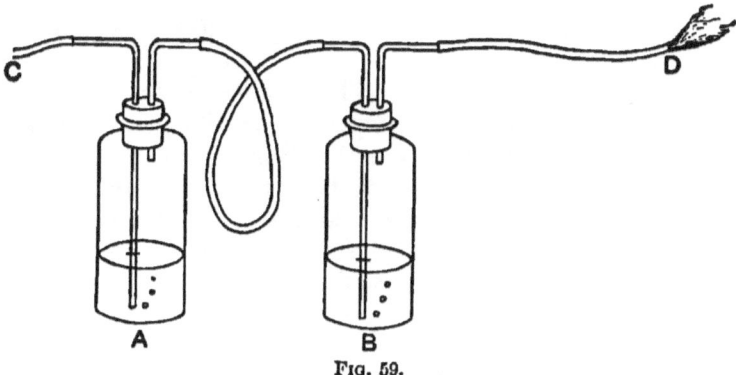

Fig. 59.

A much better way of filling a space with smoke, suitable for this and Experiment 3, is to set up two small wide-mouthed bottles, A and B in Fig. 59, and to connect them in the manner shown and easily understood. The bottle A is filled to about $\frac{1}{3}$ with strong hydrochloric acid, the bottle B with strong ammonia solution; by blowing at C, not too strongly, the bottle B is soon filled with dense white smoke which passes out at D and may be directed by the india-rubber tube into any space which is required to be filled.

Experiment 2.—Cover the open end of the tube with the mouth of a funnel of suitable size, and blow through the tube of the funnel.

Experiment 3.—The smoke-ring apparatus is very easy of construction. Bend a piece of strong pasteboard into a cylinder, about 5 or 6 inches in diameter, and 9 or 10 inches long. Glue the edges over one another; fix upon one end, with glue, a lid with a circular hole in the middle, 1 or $1\frac{1}{4}$ inch in diameter; close the other end by a piece of stout paper (as used for packing or drawing), which is drawn very tight over the end and then tied with thread. The paper must first be wetted, and stretched over the end just when the gloss arising from the wetting has disappeared. Let the apparatus get thoroughly dry for a day or two before using it.

An apparatus made of brass is better and more durable, and a piece of calf-skin or ox-bladder answers better than paper. The cylinder should have a projecting rim at the open end, so as to allow of the cover being tied firmly, otherwise it will slip off.

Experiments 4 and 5.—The fork must be one of the largest obtainable, a so-called *diapason* if possible, and the other forks to be used should also be rather large.

Experiment 6, *et seq.*—The tube should be about 15 or 20 inches long and 2 inches in diameter. The vessel must be of suitable depth for varying the depth of immersion. A tall cylindrical glass vessel is best for the purpose.

CHAPTER XIII

Experiments 1 and 2.—An ordinary bi-convex reading lens about 3 inches or more in diameter may be used for these experiments. The paper may be simply blackened with ink and used when dry. Tinfoil, as purchased from the dealers, is usually brighter on one side than on the other; the brighter surface should be used for the reflection of the rays, and the other surface should be blackened by holding it over a lighted splinter of wood which has been dipped into oil of turpentine or paraffin. To prevent the tinfoil being fused by the heat of the flame it should be rolled round a bottle filled with water, or a rather thick cylinder of metal, such as a piece of a 2-inch gaspipe.

Experiment 4.—Any tinsmith will make the canister required. It should be about 4 or 5 inches wide and 6 or 7 inches high, but only $1\frac{1}{2}$ inch in thickness. An ordinary pocket flask of metal, bright on one side and blackened on the other will do almost as well. Whilst being heated it should be placed on a rather large sheet of tin-plate which is put upon a tripod, so that the hands are protected against the heat of the burner. It is, of course, not necessary to have the canister or flask quite or nearly full of water; it is sufficient to have it $\frac{1}{4}$ full.

Experiment 5.—Two small polished discs of steel can be procured from a metalworker, and one of them can be roughened on one side by placing it for a short time into a small dish containing sufficient dilute sulphuric acid to cover the metal. When the polish has disappeared the disc is taken out, washed with clean water and dried.

Two similar knife-blades, one being made artificially rusty as described, may be used instead of the discs.

Chapter XIV

Experiments 1, 2, and 3.—The india-rubber ball should be hollow, not larger than about 1 or $1\frac{1}{2}$ inch in diameter, and if possible painted and varnished outside, so as to diminish friction. It may, with a little practice, be very easily propelled so as not to strike the sides of the blocks too much. The distance between the blocks should not be more than enough to allow the ball just to pass between them.

Experiments 4, 5, and 6.—The apparatus (Fig. 43) is made of a board cut into a semicircular shape with a saw; to the straight side a wooden rim is fixed about $2\frac{1}{2}$ inches high.

A semicircular rim of the same height is made of cardboard and fitted to the board. The diameter of the semicircle should be from 15 to 20 inches, or even more if possible. The strip of cardboard is first cut of larger size than required, placed tightly along the edge of the semicircular board, and fixed temporarily by means of a few tacks driven to about half their length through the cardboard into the semicircular edge of the board, and also at both ends of the semicircle into the side of the straight wooden rim. At each end of the semicircle a pencil line is now drawn so as to mark off precisely the length of cardboard required, since this part, from pencil mark to pencil mark, is to be divided into eighteen equal parts. The strip is then taken off the board and cut to the required length, leaving at each end, beyond the pencil line, a piece as wide as the thickness of the wooden rim, so that the end of the cardboard may be fixed upon the latter. The strip is again straightened, divided, and the divisions distinctly marked with numbers (taking care to use the ink sparingly if the cardboard is inclined to make the ink run). Apertures of about $\frac{1}{3}$ inch or a little more in diameter are then made in the proper places and the strip finally fixed permanently with glue and tacks.

The small piece of looking-glass about $\frac{3}{4}$ or $\frac{7}{8}$ inch wide, and as high as the interior of the wooden rim, should be cut with a diamond from a larger piece of looking-glass. It is fixed with a cement made by melting together in a metal spoon equal parts of resin and yellow bees' wax and stirring the mixture for a little while. When gently warmed this mixture becomes so soft that it may be moulded with the fingers. It is applied only to the edge of the mirror, not to the silvered back, which would be injured by it. While fixing the mirror the eye should be applied to the aperture at O so as to give the correct position to the mirror. It is still better to have a groove cut in the wooden rim into which the mirror fits closely.

The experiments with the apparatus are most striking when made in a dark room.

Experiments 7 and 8.—These experiments are still more striking when made before a candle flame in an otherwise dark room.

Chapter XV

Experiment 1.—For the rectangular vessel required in this and some other experiments a tin biscuit box, which must not be too large, is very suitable. It is well to paint it black inside and outside; the necessary paint may be bought in a small quantity of any "oil and colour-man." Or it may be made up by mixing 10 parts of Vegetable black, 2 of Litharge (Lead Oxide or Massicot, technically called "Dryers"), 3 parts Linseed oil, and 1 of "Turps" (spirit of turpentine). The solid materials are first powdered very finely, and thrown into the vessel in which the paint is made up; the liquids are then poured in, and the whole well stirred until the mass is of uniform consistency. The brush must be cleaned after use with oil of turpentine, or it is rendered almost useless for further work.

Experiment 2.—A very straight stout iron wire of some 8 or 10 inches in length, or a thick knitting-needle is best. One of the thick flat glass letter-weights which are sold at stationers' shops is very suitable for this experiment. The experiment may be varied by drawing a black horizontal line on a sheet of paper, and placing the glass plate upon a part of it.

Experiment 5.—Ordinary rough sandpaper will do very well. It should be bent at the edges of the board, and fastened well at the back or lower side of the board by means of numerous short tacks. The slanting margin especially must lie in close contact with the board, or the pencil is apt to stop. The board should be at least a yard long, and a foot or a foot and a half wide. Its inclination should be such as to render the motion slow enough for observation, and yet not too sluggish, or the pencil may be brought to rest altogether at the edge of the rougher surface.

Chapter XVI

Experiment 1.—For this experiment the aperture through which the rays are to pass must be extremely small. It is best to cut neatly, in a sheet of cardboard, a round hole about 1 inch in diameter, which can be afterwards used for the further experiments. For this experiment paste a small sheet of tinfoil over the hole, pressing it down smoothly, and when dry prick a hole in the tinfoil with a pin.

The experiments in which images are to be observed may in many cases be so arranged that the image is thrown upon a wall or any vertical flat surface. A movable screen is, however, much better, and saves a great deal of labour. It is best to have a frame made by a joiner for the purpose and to stretch paper over it. The frame should be between 2 and 3 feet high and about the same breadth, and should be made of wooden laths, 1 inch wide, and about one-third of an inch thick. Common white drawing paper may be used, but if it is intended to show the experiments to a large audience, white tissue paper is better, for it is translucent and the images may be seen on either side of the screen. If drawing paper is used it must be first laid upon a table and damped with a clean moist sponge or cloth before stretching it upon the frame; the frame is then covered on one side with a layer of glue, pressed upon the moist paper, raised with the latter, and turned over; the paper should then be well pressed upon the frame. When dry the paper will be tight and smooth. Tissue paper cannot be moistened without tearing it; it hardly bears the moisture of the glue or gum, and it should be fixed by means of a very thin layer of Canada Balsam spread upon the frame with the fingers, so as to remove creases as much as possible. The Canada Balsam must be allowed to dry a little upon the frame, if it is very thin; otherwise the paper will not adhere to the frame when the latter is raised. The fingers may be cleaned from the Balsam with oil of turpentine.

It is very convenient to have a proper stand, into which the frame of the screen can be clamped, made at the same time; or the screen may be held in an upright position by placing it between two of the stouter blocks which have been used in previous experiments, and which for the present purpose must be laid with their flat sides upon the table.

Experiment 3.—The triangular vessel is made of three rectangles of glass, cut neatly with a diamond out of a sheet of window glass. Each rectangle is 5 inches long and $1\frac{1}{2}$ inch wide. The long sides are joined together carefully along their whole length by sealing-wax, which will only adhere well if the glass is hot while the sealing-wax is applied. The little vessel must of course be watertight. Two triangular pieces of wood cut to the required size from a thin lath are fixed to the ends, one end being first closed, the prism filled with liquid, and then the other end closed. It is of course not necessary to have a third, upper, side to the vessel at all; joining the edges of two glass rectangles and attaching the wooden ends, produces a kind of small trough which, in the experiments with water, will be sufficient if it is held with its joined edges downwards. But on account of the unpleasant smell of the Carbon Disulphide and its rapid evaporation it is very desirable to have the prismatic vessel which contains it well closed everywhere. Carbon Disulphide dissolves sealing-wax, hence the experiment should not be undertaken unless a thick layer of sealing-wax has been laid on everywhere along the edges; and after the experiment the vessel should be taken to pieces again, *not* by heating the sealing-wax, which might give rise to most serious accidents, as Carbon Disulphide and its vapour are highly inflammable bodies, but by holding the vessel perpendicularly, scraping off with a knife the sealing-wax along the upper end, and then pouring the Carbon Disulphide through a funnel into a bottle provided with a well-ground glass stopper.

Triangular glass bottles for these experiments, carefully ground and polished, may be bought, but they are somewhat expensive.

A strong solution of Lead Acetate (Sugar of Lead) in water, to which a little acetic acid is added if it should be milky, will serve very well instead of Carbon Disulphide to show a greater dispersion than that of water. For this solution the same little vessel which held the water may without any inconvenience be used.

It is necessary in all these experiments to support the prismatic vessel, or prism used, in a fixed position, so as to have one's hands free for the adjustment of the screen, aperture, etc. It is best to bend two copper wires making two angles near the ends equal to the angle of the prism. The prism is put into these angles near each end of it, and the other ends of the wires are bent into hooks and suspended from a suitable support.

Experiment 5.—Ruby red and cobalt blue glasses are best to use. Small plates of these two kinds are easily obtained from dealers in chemical apparatus. The plates should be as thin as possible especially the blue, or too little light will pass through them.

Experiment 9.—This experiment requires a little manipulation of the glass plate C, which must be moved while the eye looks obliquely through it until the reflected image of one paper exactly coincides with the other paper.

Chapter XVII

Experiments 1 and 2.—A few grains of either substance should be allowed to dissolve in an ounce of water in small beaker glasses, gently stirred, and when dissolved a portion of each transferred to two test-tubes. No undissolved residue should be poured with the liquid into any of the test-tubes. The salt solution should be dropped slowly into the nitrate solution as long as the precipitate increases.

Before proceeding to Experiment 2 the liquid should be poured off from the precipitate, water should be added and then carefully poured off again; this "washing" of the precipitate should be repeated with fresh water several times.

Experiment 3.—If possible, direct sunlight should be allowed to fall upon one side of the test-tube, which may for that purpose be placed into some corner which will sufficiently protect the other side.

Experiment 4.—The necessary steps for producing a photographic image so as to illustrate the practical procedure in photography are comprised in the following successive operations:

(a) Obtain from any dealer in photographic chemicals a small quantity of iodised collodion. This is generally a solution of Potassium Iodide, and Bromide, in a mixture of Alcohol and Ether, which also holds a body resembling gun-cotton (pyroxiline) in solution. The Iodides and Bromides of Silver are chiefly employed in photography because they are far more sensitive than the Chloride, which for the sake of its greater facility of preparation has been used for the experiments described in the text. Pour some of this liquid over a small and clean glass plate; the ether which acted as solvent evaporates, and a transparent film is left on the glass which includes the Iodide and Bromide of Potassium.

(b) Before this film is quite dry bring the glass into a room from which daylight is quite excluded, and which is lighted only by a candle flame surrounded by a screen of yellow or orange tissue-paper (obtainable at the photographic dealers). Lay the glass with the side bearing the film upwards on a shallow tray or dish, and pour into the latter (but not directly on the film) a solution of Silver Nitrate (about 30 grains of Nitrate dissolved in 1 oz. of pure water and previously filtered), so that the liquid may cover the collodion film. In consequence of the mutual decomposition, as in the case of the common salt, the film loses its transparency and becomes yellowish from the formation of iodide of silver. The plate is now called the *sensitive* plate. Remove it and allow it to drain.

(c) Cut some patterns, letters, or a star, out of a piece of cardboard, place it over the moist film so that it may not actually touch it, and hold it for fifteen or twenty seconds at the open door of the room in diffused daylight, then bring it back into the dark room and remove the cardboard.

(d) No image will yet be visible on the film, unless it has been exposed too long. Place the glass into a clean dish with the film uppermost, and pour over it a solution of green vitriol (Ferrous Sulphate), which contains about 15 grains in 1 ounce of water, and to which a few drops of acetic acid have been added. The film will now gradually darken until the form of the image is quite distinct. The Ferrous Sulphate is in this action technically called a "developer"; it is a powerful agent for withdrawing oxygen, and it develops the chemical change which the light has produced in the first instance into a marked chemical difference between the exposed part of the film and the unexposed portion.

(e) The plate is now dipped into a strong solution of Sodium Hyposulphite, which removes the Silver Iodide which has been unacted upon; otherwise this would afterwards be acted on by the light, and would form together with the image a uniform black surface. The "Hypo" is the "fixing" agent.

(*f*) The film is now well washed, and can be exposed to the light. The image consists of reduced silver on a clear ground, and is called the "Negative." From it, by acting on prepared paper, "Positives" may be obtained; this operation is called "printing." The paper may either be bought, or prepared by soaking paper in a solution of common salt, and after drying it dipping it in a Nitrate of Silver solution in the dark. When dry the negative is placed upon it and the paper exposed to the light; gradual discoloration will take place wherever it is not protected by the dark portions of the negative. The positive is "fixed" like the negative by immersion in Hyposulphite solution.

CHAPTER XVIII

Experiment 1, *et seq.* — An ordinary glass tumbler is sufficiently capacious for these experiments. The strips of zinc and copper should be cut into rectangles of about 5 or 6 inches in length, so as to project about an inch beyond the mouth of the tumbler, and 1 or $1\frac{1}{2}$ inch in width. For the first experiment zinc and copper *foil* should be used, so as to observe more easily the gradual wasting away of the zinc. For the remaining experiments the strips may be cut from ordinary sheet-zinc and sheet-copper. If there should be any difficulty, from the shape of the tumbler, in keeping the strips apart, two narrow pieces of cork or wood, of suitable size, may have a slit cut into each to a small depth for receiving the immersed ends of the strips; this will effectually prevent their coming into contact with one another during the experiments. It is best to have the copper wire neither too fine nor too thick, about No. 36; it should be covered with gutta-percha or india-rubber for most electrical experiments. If such wire is not to be had, ordinary uncovered wire may be used for most experiments, especially if a few inches from the ends small lumps of sealing-wax are fused round the wire which serve for handling them with the fingers. The ends themselves must invariably be very bright, and the metallic surfaces which are to be connected must frequently be scraped with a knife. The ends of the wires which are to be fixed to the plates are best attached by means of so-called "binding-screws" or clamps, which may be had in a great variety of forms from the dealers in electrical apparatus.

Experiments 9 and 10.—The nail will easily attract small iron nails, keys, etc., while the current passes round it, and these when the current is broken drop off more quickly than iron filings. For magnetisation the wire wrapped round the iron must invariably be covered with gutta-percha or some other "insulating" substance.

Experiment 11.—It is best to clamp the wire through which the current is to pass in the proper position, about 1 inch above the magnet and parallel to it, before making connection.

A magnetised knitting-needle, suspended by a silk thread in a horizontal position, answers very well, but it must not be forgotten that the current from one voltaic couple will give an appreciable deflection only when the magnet is extremely carefully poised. The little battery, described in Chapter XIX. will produce a very good deflection, especially soon after it has been freshly filled.

The connecting wires, especially for the last three experiments, should be each several yards long.

Chapter XIX

Experiment 1.—The little battery is built up of six ordinary test-tubes, which are placed side by side in holes of suitable diameter bored in a pretty stout piece of board. Six strips of copper and six of zinc are cut out of thin sheet-metal, each about 3 or 4 inches long and $\frac{1}{4}$ inch wide; five of the strips of zinc are soldered each of them to one of the strips of copper, allowing the ends to overlap each other for about $\frac{1}{6}$ or $\frac{1}{8}$ of an inch. The sixth strip of each metal is left free, and these form respectively the poles of the battery; each of them has a thin copper wire, several feet long, soldered to it. It is best to use wire covered with silk or gutta-percha. The joined strips are bent as shown in Fig. 54, and each test-tube must contain a strip of each metal. The two single strips form the terminals, one being placed in the first and the other in the last test-tube. The two different metals in each test-tube must not be allowed to touch each other; their contact is prevented by fixing small corks between the metals, as shown in the figure. These corks should only fit loosely, otherwise, if pressure is required in pushing them between the metals the thin test-tubes will probably be broken. The liquid used should be introduced very carefully by means of a small funnel, so as not to moisten the outsides of the test-tubes.

Solder is prepared by melting 3 parts by weight of tin with 2 parts by weight of lead in an iron ladle. The mixture is stirred and then poured on a flat horizontal surface, such as an old plank or a flat stone, that it may form a thin cake from which small pieces may be cut with a strong knife or a pair of cutting pliers. In applying solder it is essential that the surfaces to be united be quite free from oxide, which would prevent adhesion of the solder; this is insured by sprinkling a little powdered borax upon the surfaces to be joined. A small pellet of solder is then placed upon one surface, the other pressed upon it, and the part held over the point of a flame until the solder runs, when the part is immediately removed from the flame, and the surfaces will, when cold, be found to adhere.

Chapter XX

Experiments 1 and 2.—It is essential in these experiments that the liquid used for each couple should be of uniform strength. A mixture of 1 part of sulphuric acid with 10 parts of water should therefore be made in sufficient quantity to supply each different couple with some of the same liquid, and the effect upon the magnet should be in each case observed at once. The liquid from each cell may be poured into a bottle and made use of afterwards for ordinary battery purposes.

Plates of the different metals, including a carbon plate, all of the same size as the zinc plate used, may be obtained from dealers in electrical apparatus; they are all very inexpensive except the platinum plate, which may be omitted from the experiments where the avoiding of expense is a matter of importance.

Experiment 4.—The ends of the platinum wire may be simply twisted several times round the ends of the two terminals from the battery, so that the platinum wire forms a link between them. This must of course be done before the battery is in action. It is best to keep one of the wires clamped by the binding-screw to the zinc, or carbon, of one battery cell, and unclamp and remove the other; when the platinum wire is firmly fixed, the loosened clamp is again used for fixing the wire to the battery pole from which it had been removed.

A Bunsen's battery of four or five cells will be indispensable for some of the experiments on the action of the current, though for this and the next experiments in the present chapter one or two cells are sufficient.

In a Bunsen cell the positive pole is formed by a four-sided prism of dense carbon of good conductivity; the negative pole by a hollow cylinder of zinc. The two bodies which thus form the opposite poles are immersed in two different liquids which are placed in conducting communication by means of a thin cylindrical vessel of unglazed earthenware, which permits conduction through its substance, as this becomes saturated with the liquids on either side, but yet prevents them from mixing with one another. An outer vessel of glass or glazed stoneware encloses the whole tube. Into this is put the zinc cylinder, within this the "porous vessel," and within the porous vessel the carbon. To about twenty parts by volume of water one part of sulphuric acid is added; this forms the liquid which is to be poured round the zinc by means of a funnel while the porous vessel is held in its place by pressing the finger upon it, otherwise it will float, and it is then impossible to see whether the proper quantity of dilute sulphuric acid has been poured into the outer vessel, which must appear to be full to within an inch from the top. The porous cell is then filled, while the carbon is in it, with commercial concentrated nitric acid, pouring it in by means of a small funnel, very carefully avoiding the dropping of any of it on the zinc, binding-screws, and connecting wires, or into the sulphuric acid; the funnel should therefore not be removed till the last drop of the acid has run out, and then its end should be held close to the inner edge of the porous vessel for a short time to make sure that no drop of acid is adhering to it. The acid not only destroys the metal parts, but also the amalgamation, if it mixes with the sulphuric acid.

The zinc cylinders obtained from the dealers are usually amalgamated. If this is not the case, or the amalgamation should have become defective by long disuse of the cells, the zinc may easily be amalgamated by dipping it first for a short time in dilute sulphuric acid, and then holding it over a capacious pan, and pouring a few drops of mercury upon the metal. If the acid has produced a clear surface upon the zinc the mercury will spread itself out upon it at once; otherwise it is well rubbed over the surface everywhere, inside and outside the cylinder, by means of a little piece of cloth moistened with dilute acid. The mercury left should not be poured back into a bottle containing pure mercury, as it is alloyed with zinc.

The terminal binding-screw of the carbon is a piece of brass bent twice at right angles; sometimes a screw attached to one side allows it to be fixed direct to the carbon, sometimes this is more firmly done by means of a small movable plate which has the spindle of another screw on the top of it attached to it, which slides in a slot, so that the plate can be pushed to and fro by the lateral screw when the top screw is loosened, and then firmly clamped when in the proper position. The zinc has a simple binding-screw attached to it with a hole, and an additional screw for receiving and clamping the wire which forms the negative terminal. The carbon binding-screw has a similar hole and screw for the positive terminal. If two or more cells are to be united, the carbon of the first cell is to be connected by a short thick wire with the zinc of the second, the carbon of the second with the zinc of the third, and so on; the longer wires which form the two terminals are attached to the zinc of the first cell and the carbon of the last respectively, each wire having one end free.

While the battery is in use red fumes of nitrous acid are constantly evolved, especially when the nitric acid is no longer new. These vapours are very suffocating and unpleasant. It is best to place the battery outside the room upon a ledge of the window if possible, and lead the terminal wires into the room by two small holes in the sash, or by means of flat strips of copper which may be easily pushed between the sash and the window, and to which at their opposite ends the wires may be soldered. The production of vapours may be almost entirely checked by saturating the nitric acid before using it with Nitrate of Ammonium.

In taking the battery to pieces the carbon screws are removed first, then the carbon lifted to the edge of the porous pot, and held there till the acid has run off. A pan of water is kept ready, and the carbon placed into it. The porous pot is next lifted out and the acid poured into a bottle; the empty pot is then also transferred to the water. Last of all, the zinc is removed. Every part should now be rinsed in a stream of water from a tap, and the zinc and screws wiped dry, while the porous cells and carbons had better be kept in water for several days, changing the water from time to time. Finally, they are taken out and left to dry. If after a little time the porous vessels show an incrustation of white fur, or the carbons should smell of nitric acid, these pieces must again be placed in water for some time.

For many experiments a small apparatus for quickly and easily opening and closing a circuit, a so-called "contact breaker" is very useful, as for example in Experiment 4, when it will save the trouble of detaching the screw at one pole while the platinum wire is introduced. It is very inexpensive, and may be obtained at any dealers in electrical appliances.

Experiment 6.—Small carbon rods for this experiment are obtainable at the dealers: they are very inexpensive.

Chapter XXI

Experiment 1, *et seq.*—Litmus paper may be purchased from dealers in chemicals. A solution of Litmus may be prepared by purchasing solid Litmus and boiling a small quantity of it with distilled water in a beaker glass until the liquid assumes a deep purple tint. It may be replaced by the more accessible colouring matter contained in most blue flowers—for example, the violet, lobelia, larkspur, and the common blue flag (*Iris Germanica*) of our gardens. All these yield blue pigments which are changed to red by acids, and *green* by alkalies, as soda, etc. To obtain the colouring matter some of the flowers should be placed into a china dish with some alcohol. The dish is placed upon the mouth of a beaker glass which is one-third or half full of water, and the latter is gently heated over a burner so as to prevent too rapid boiling. After a quarter of an hour's boiling the dish is removed, the flowers placed in a little muslin, which is formed into a small bag, and the colouring matter squeezed into a clean china dish. This is again placed over the "water-bath," and the water in the beaker heated until in the dish a blue pigment is left. This is soluble in water, and a small quantity of it may be used for colouring the solution of sulphate of soda in Experiment 8.

When "red cabbage" is obtainable, a very easy plan is to cut up a few leaves, just cover them with water in a beaker glass, and heat the water until it *begins* to boil. The liquid is then poured through a filter and a small quantity of sulphate of soda dissolved in it (about 20 or 25 grains in an ounce of the blue solution).

Experiment 5.—For this experiment a solution of about an ounce of soda (not washing soda but *caustic*) in a pint of water, and a similar quantity of sulphuric acid in half a pint of water, should be prepared beforehand, and each separately well mixed by shaking. The soda should be dissolved by heating the water. An ounce accurately measured of the soda solution should be placed in a beaker and the dilute sulphuric acid *dropped* into it very slowly, if possible, from a graduated pipette (or burette), at any rate from an ounce measure, which is graduated into small divisions, so that we are able to know exactly how much of the acid has gone to neutralise an ounce of the basic solution. In that exact proportion they must be mixed afterwards for the formation of the sodium sulphate to be obtained by evaporation. The coloured solution made in Experiment 5 will, of course, after evaporation, give the salt, but it will be covered with coloured specks, which will scarcely allow a good observation of the character of the pure salt. On the other hand, when a mere random mixture of acid and base is made we obtain the salt, but it is very difficult, especially for beginners, to free it from the excess of alkali or acid which clings to it and renders its application impossible for our purpose. If there should be any difficulty in the measuring out we may help ourselves thus : Evaporate the coloured solution obtained in Experiment 5 ; transfer the dirty-looking salt obtained to a little thin china crucible, and heat over a strong Bunsen flame for some time until the mass is red hot. The colouring matter is thus burnt and carbonised. Let the whole cool and then throw it into water; boil and filter. A colourless pure solution of sodium sulphate will run through which may be evaporated in a dish and will yield the pure salt. The particles of carbon will remain on the filter.

Experiment 8.—The little vessel is easily made by bending a piece of glass tubing about 6 inches long and $\frac{1}{4}$ inch wide into the required shape. It can easily be made to stand upright by sticking it into a flat piece of wax or sealing-wax, or even a large bung, into which a suitable slit is made with a knife.

The poles for most chemical decompositions by the current, especially when acids are separated, must be formed of platinum foil, because all metals except gold and platinum are attacked by some of the products of the decomposition, and the result would thus be vitiated.

The two platinum poles should be cut so as to be about 2 inches long, and of a width which will allow them easily to pass in and out of the little vessel. They are either soldered to bits of platinum wire, the free ends of which are firmly twisted together with the ends of the copper wires from the battery, or they may be attached thus : make a pin hole near one end of the platinum foil, bend the end of the platinum wire into a small hook, pass the end of this through the pin hole, and when the foil hangs on the hook close the hook by the pressure of the fingers ; finally, place the platinum foil and wire upon some hard surface, and fasten wire and foil firmly together by one or two smart blows with a light hammer.

Experiment 10.—Special apparatus for electrical decompositions, involving more or less expense, are sold by the dealers ; but it is quite possible to demonstrate the principal facts of electrolysis, for ordinary school purposes, by means of home-made apparatus. The two test-tubes shown in Fig. 58

must of course be clamped in suitable stands, so-called retort-stands, which are quite indispensable for experimental work.

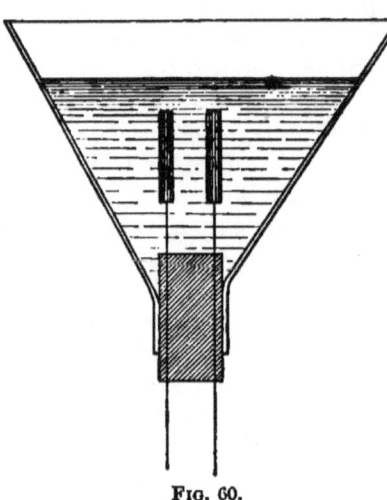

Fig. 60.

These are provided with convenient rings and clamps for such purposes. Test-tubes are also sold with little glass-hooks or glass-rings fixed to their closed ends, so that they may be hung to some support over the platinum strips. A very convenient way of carrying out this experiment is to cut off the tube of a large wide funnel, about one inch and a half from where it joins the wider part; pass a well-fitting cork through the remainder of the tube, and pass the wires which carry the platinum strips through small holes made in the cork and fasten them with a little sealing-wax. The test-tubes will easily stand without further support on the cork itself; the vessel may be very conveniently supported on a wide ring attached to the retort-stand. Fig. 60 will explain the arrangement.

Chapter XXII

Experiment 1.—About 50 grains of copper sulphate dissolved in $1\frac{1}{2}$ ounce of water are sufficient for this experiment.

Though the small V-tube used for the decomposition of the sodium sulphate is suitable for exhibiting the results of the present decomposition completely, yet it is convenient to carry out the experiment in a somewhat larger V-tube, which can easily be made from a piece of soft glass-tubing about $1\frac{1}{2}$ or 2 feet long and an inch or a little less in diameter. It should be bent in the middle over an ordinary fan-tail burner, heating the tube well *along* not across the flame until the heated part is soft, and it can *easily* be bent so that the halves make an angle of about 40 degrees. Platinum wires with attached strips can easily be fixed by passing the ends of the wires through corks which fit the bore of the tube, and bending them outside into a little hook so as to prevent their slipping back. These little hooks are very convenient for making connection with the battery wires. Each cork should have a small notch cut out at the side, so as to allow those gases which are produced in the decomposition to escape freely. Special V-tubes for electrical decompositions with suitable stands are sold at the dealers; they are very convenient, but somewhat expensive.

Experiment 3.—The strips should be about 2 inches square, and the decomposition carried out in a small but wide jar, which will allow the strips to be completely immersed at a distance of about 1 inch from one another.

Tin chloride may be made by dissolving a few scraps of granulated tin (that is, tin melted in a spoon and poured in a fine stream from a height of 3 or 4 feet into a pail of water) in hot hydrochloric acid, and adding water to it.

Silver nitrate when dissolved in ordinary water gives a cloudy solution, which must be filtered several times if necessary. Distilled water will give no cloudiness, which is due to the soluble chlorides, always present in ordinary water, producing a white precipitate of chloride of silver.

Lead Acetate (Sugar of Lead) also gives a cloudy solution; this may be rendered clear by adding a few drops of acetic acid (white vinegar) or nitric acid.

THE END

Printed by R. & R. CLARK, *Edinburgh*

www.ingramcontent.com/pod-product-compliance
Lightning Source LLC
Chambersburg PA
CBHW032142230426
43672CB00011B/2421